ENERGY POLICIES, POLITICS AND PRICES

ENERGY EFFICIENCY FINANCING PROGRAMS

RATIONALES, OPTIONS, AND LIMITS

ENERGY POLICIES, POLITICS AND PRICES

Additional books in this series can be found on Nova's website under the Series tab.

Additional E-books in this series can be found on Nova's website under the E-book tab.

ENERGY POLICIES, POLITICS AND PRICES

ENERGY EFFICIENCY FINANCING PROGRAMS

RATIONALES, OPTIONS, AND LIMITS

**LOUISE ALTMAN
EDITOR**

New York

Copyright © 2014 by Nova Science Publishers, Inc.

All rights reserved. No part of this book may be reproduced, stored in a retrieval system or transmitted in any form or by any means: electronic, electrostatic, magnetic, tape, mechanical photocopying, recording or otherwise without the written permission of the Publisher.

For permission to use material from this book please contact us:
Telephone 631-231-7269; Fax 631-231-8175
Web Site: http://www.novapublishers.com

NOTICE TO THE READER

The Publisher has taken reasonable care in the preparation of this book, but makes no expressed or implied warranty of any kind and assumes no responsibility for any errors or omissions. No liability is assumed for incidental or consequential damages in connection with or arising out of information contained in this book. The Publisher shall not be liable for any special, consequential, or exemplary damages resulting, in whole or in part, from the readers' use of, or reliance upon, this material. Any parts of this book based on government reports are so indicated and copyright is claimed for those parts to the extent applicable to compilations of such works.

Independent verification should be sought for any data, advice or recommendations contained in this book. In addition, no responsibility is assumed by the publisher for any injury and/or damage to persons or property arising from any methods, products, instructions, ideas or otherwise contained in this publication.

This publication is designed to provide accurate and authoritative information with regard to the subject matter covered herein. It is sold with the clear understanding that the Publisher is not engaged in rendering legal or any other professional services. If legal or any other expert assistance is required, the services of a competent person should be sought. FROM A DECLARATION OF PARTICIPANTS JOINTLY ADOPTED BY A COMMITTEE OF THE AMERICAN BAR ASSOCIATION AND A COMMITTEE OF PUBLISHERS.

Additional color graphics may be available in the e-book version of this book.

Library of Congress Cataloging-in-Publication Data

ISBN: 978-1-63117-200-7

Published by Nova Science Publishers, Inc. † New York

Contents

Preface		vii
Chapter 1	Getting the Biggest Bang for the Buck: Exploring the Rationales and Design Options for Energy Efficiency Financing Programs *Mark Zimring, Merrian Borgeson, Annika Todd and Charles Goldman*	1
Chapter 2	Scaling Energy Efficiency in the Heart of the Residential Market: Increasing Middle America's Access to Capital for Energy Improvements *Mark Zimring, Merrian Borgeson, Ian M. Hoffman, Charles A. Goldman, Elizabeth Stuart, Annika Todd and Megan A. Billingsley*	53
Chapter 3	The Limits of Financing for Energy Efficiency *Merrian Borgeson, Mark Zimring and Charles Goldman*	75
Index		91

PREFACE

Many state policymakers and utility regulators have established aggressive energy efficiency (EE) savings targets which will necessitate investing billions of dollars in existing buildings – and tax payer and utility bill payer funding is a small fraction of the total investment needed. Given this challenge, some EE program administrators are exploring ways to increase their reliance on financing with the aim of amplifying the impact of limited program monies. This book explores the rationales and design options for energy efficiency financing programs; discusses the increasing middle America's access to capital for energy improvements; and provides insight on the limits of financing for energy efficiency.

Chapter 1 – Many state policymakers and utility regulators have established aggressive energy efficiency (EE) savings targets which will necessitate investing billions of dollars in existing buildings – and tax payer and utility bill payer funding is a small fraction of the total investment needed. Given this challenge, some EE program administrators are exploring ways to increase their reliance on financing with the aim of amplifying the impact of limited program monies. While financing is potentially an attractive tool for increasing program leverage and mitigating the rate impacts of utility customer-funded efficiency programs, administrators can face difficult choices between allocating funds to financing or to other approaches designed to overcome a broader set of barriers to consumer investment in EE. Robust assessments of financing's role in reducing energy use in buildings are necessary to help policymakers and program administrators make better choices about how to allocate limited resources to achieve cost- effective energy savings at scale.

Chapter 2 – Middle income American households – broadly defined here as the middle third of U.S. households by income – are struggling. Energy improvements have the potential to provide significant benefits to these households – by lowering bills, increasing the integrity of their homes, improving their health and comfort, and reducing their exposure to volatile, and rising, energy prices. Middle income households are also responsible for a third of U.S. residential energy use, suggesting that increasing the energy efficiency of their homes is important to deliver public benefits such as reducing power system costs, easing congestion on the grid, and avoiding emissions of greenhouse gases and other pollutants.

While middle income Americans have historically invested in improvements that maintain and increase the value of their homes, they have seen an important source of financing – the equity in their properties – evaporate at the same time that their access to other loan products has been restricted. A number of energy efficiency programs are deploying credit enhancements, novel underwriting criteria, and innovative financing tools to reduce risks for both financiers and borrowers in an effort to increase the availability of energy efficiency financing for middle income households. While many of these programs are income-targeted, the challenges, opportunities, and emerging models for providing access to capital may apply more broadly across income groups in the residential sector.

Chapter 3 – Financing is an appealing concept when efficiency program budgets are a small fraction of the overall level of efficiency investment needed to achieve our public policy goals – but that does not mean financing is always the solution, and it is certainly not the *only* solution. The authors show that financing can, in some cases, increase the leverage of public dollars. In most cases, however, it is not able to drive demand to the same degree as direct incentives like rebates and so cannot be expected to replace other incentives in the current marketplace. The authors also show that subsidized financing for those who already have access to capital may be a poor use of public funds, and that increasing access for those who are currently underserved will likely require ongoing subsidy. This is not to say that financing is unimportant – financing is one of many important tools for scaling efficiency and should be employed thoughtfully with the questions outlined in this paper in mind.

In: Energy Efficiency Financing Programs
Editor: Louise Altman

ISBN: 978-1-63117-200-7
© 2014 Nova Science Publishers, Inc.

Chapter 1

GETTING THE BIGGEST BANG FOR THE BUCK: EXPLORING THE RATIONALES AND DESIGN OPTIONS FOR ENERGY EFFICIENCY FINANCING PROGRAMS[*]

Mark Zimring, Merrian Borgeson, Annika Todd and Charles Goldman

EXECUTIVE SUMMARY

Many state policymakers and utility regulators have established aggressive energy efficiency (EE) savings targets which will necessitate investing billions of dollars in existing buildings – and tax payer and utility bill payer funding is a small fraction of the total investment needed.[1] Given this challenge, some EE program administrators are exploring ways to increase their reliance on financing with the aim of amplifying the impact of limited program monies.[2] While financing is potentially an attractive tool for increasing program leverage and mitigating the rate impacts of utility customer-funded efficiency programs, administrators can face difficult choices between allocating funds to financing or to other approaches designed to overcome a broader set of

---

[*] This is an edited, reformatted and augmented version of the Lawrence Berkeley National Laboratory and prepared under contract for the U.S. Department of Energy, dated December 2013.

barriers to consumer investment in EE. Robust assessments of financing's role in reducing energy use in buildings are necessary to help policymakers and program administrators make better choices about how to allocate limited resources to achieve cost- effective energy savings at scale.

In order to better understand what EE financing can be reasonably expected to achieve, and for whom, this report is organized around three levels of inquiry (Figure 1), from the most fundamental (level 1) to the most detailed (level 3).

For each of these three levels of inquiry, the report describes key uncertainties that must be resolved in order to better understand the role of financing in delivering cost-effective energy savings. Examples include:

- What market segments or types of efficiency improvements are currently underserved by financial markets and why?
- Is financing an effective tool for driving consumer EE adoption? For which consumers and at what cost? What other strategies should be combined with financing to maximally increase EE adoption at the lowest possible cost?
- Does financing for EE have lower participant defaults and delinquencies than financing for other property improvements? If so, is the default rate low enough to warrant substantial improvements to the interest rates, loan lengths and/or underwriting for private financial products? What impact do these improved features have on consumer EE adoption?
- Does sufficient consumer demand exist today to warrant program investments in aggregation and securitization infrastructure, or should interventions simply focus on increasing the volume of standardized financial products?
- Are novel financing products more effective in overcoming barriers to EE adoption than traditional financing products?
- Are consumers as (or more) likely to adopt targeted EE improvements if offered financing or rebates (or other support such as technical assistance)? Do completed EE projects deliver greater energy savings if program financing is used (or available) compared to rebates (or other strategies)?

This report offers a starting place for developing a better understanding of financing's role in driving cost- effective EE adoption. We encourage program administrators and policymakers to identify those issues and questions that are most relevant to their program's success and to begin to test whether their assumptions are correct. Not every program needs to answer every question – as more and more programs actively explore these questions, lessons learned can be shared.

Chapter 1. Introduction

Many state policymakers and utility regulators have established aggressive energy efficiency (EE) savings targets which will necessitate investing billions of dollars in existing buildings – and tax payer and utility bill payer funding is a small fraction of the total investment needed.[3] Given this challenge, some EE program administrators are exploring ways to increase their reliance on financing with the aim of amplifying the impact of limited program monies.[4] While financing is potentially an attractive tool for increasing program leverage and mitigating the rate impacts of utility customer-funded EE programs, it is critical that policymakers and program administrators gain a better understanding of what financing can be reasonably expected to achieve, and for whom – and how to design financing programs to both maximize short-term impacts and to learn from experience over time.

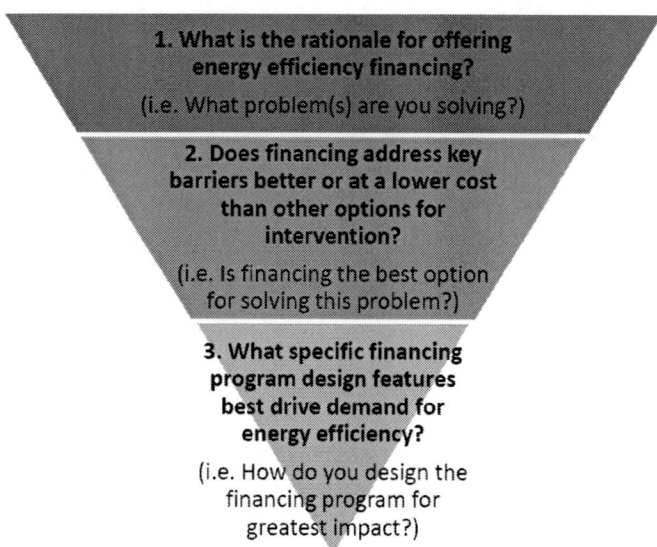

Figure 1. Three levels of inquiry to inform the design of energy efficiency financing programs.

Background

Financing has historically been a small part of the portfolio of energy efficiency program offerings. Because these initiatives have been small and often secondary to rebate and other programs, the efficacy of financing programs in delivering cost-effective energy savings has typically been assessed qualitatively (Cadmus 2012), or not at all (beyond simply tracking the financing amounts issued). In most cases, these initiatives have failed to achieve significant market penetration (Fuller 2009, Hayes et al. 2011). In a world of limited program budgets, administrators can face difficult choices between allocating funds to financing or to other approaches designed to overcome a broader set of barriers to consumer investment in EE.

As some policymakers and program administrators consider shifting the traditional mix of program offerings to rely more heavily on financing, it is important to undertake a more rigorous assessment of the ability of financing to overcome barriers to consumer adoption of property improvements that deliver cost- effective incremental energy savings – and be able to compare the impacts of investments in financing programs (e.g., cost and level of energy savings, rate impacts) to other programmatic strategies. Robust assessments of financing's role in reducing energy use in buildings will help policymakers and program administrators make better choices about how to allocate limited tax payer and utility bill payer resources.

Report Objectives

The primary objectives of this report are to articulate the rationales for offering financing programs, to highlight key policy and program design questions regarding the role of financing for which we need better answers, and to offer guidance to administrators on how financing programs can be designed and evaluated to assess their efficacy. Some of these questions can be tested directly by assessing the impacts of specific program designs; other questions will require more qualitative market research, and observation over time. We divide these questions into three "levels" of inquiry, represented in Figure 2, from the most fundamental (level 1) to the most detailed (level 3).

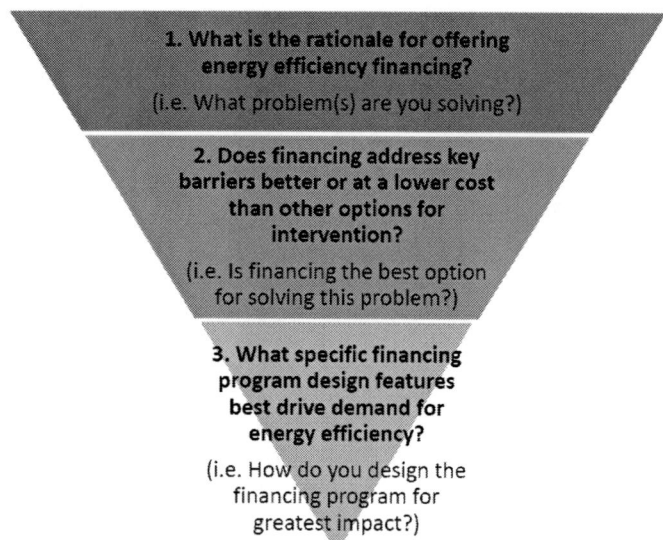

Figure 2. Levels of inquiry to inform the design of energy efficiency financing programs.

Report Organization

Level 1, described in the next chapter, explores several possible rationales for devoting tax payer or utility bill payer funds to efficiency financing programs; understanding the "problem" that financing is intended to address is vital to tracking and evaluating the ability of a program to effectively address barrier(s) to increased EE adoption. Level 2, discussed in Chapter 3, explores the efficacy of financing *relative* to the many other options for market intervention. Level 3, described in Chapter 4, explores key questions that must be resolved in designing effective financing programs once a program administrator determines that support for financing is needed.

In Chapter 5, we briefly describe several methods that can be utilized to address these questions: (a) qualitative market research that could include market assessments and participant surveys, (b) the analysis of standardized financing program data, ideally compared across multiple jurisdictions, and (c) the use of experiments to more definitively assess program impacts and the efficacy of program design features. In Appendix A, we offer a more detailed description of experimental design techniques and examples of how these techniques can be used to answer key financing questions.

How to Read This Report

This report provides an overview of the broad motivations for offering financing initiatives that facilitate efficiency investments across consumer segments as well as key questions that must be resolved to assess the efficacy of these strategies. Rather than attempt an exhaustive catalog of the range of uncertainties about the impacts of these programs on delivering incremental energy savings, we offer illustrative examples of those questions that we believe are most important to answer based on our experience working with EE financing program administrators, policymakers, financial institutions and EE service providers. It is essential to ask the key questions identified in this report with an appreciation that both their importance and their answers may differ substantially between, and within, consumer classes depending upon local market conditions, targeted improvements and policymakers and program administrator goals. We encourage stakeholders to identify the issues and uncertainties most relevant for their target markets and design future programs to include thought out plans for answering key program impact and design issues.

CHAPTER 2: WHAT IS THE RATIONALE FOR OFFERING A FINANCING PROGRAM?

At the most fundamental level, the rationale(s) for offering financing programs must be clearly established so that program administrators have an understanding of the problem(s) they are aiming to solve, and can recognize "success" when (and if) it occurs. Many state policymakers and utility regulators acknowledge that EE has substantial public and energy system benefits. However, consumers often invest in EE at lower levels than is socially-optimal given the combination of public, energy system, and private benefits (Golove & Eto 1996, Jaffe & Stavins 1994). A variety of barriers to broader consumer EE adoption have been cited in the literature, including the fact that EE often has "high first costs" (IEA 2008; Jaffe & Stavins 1994). While these up-front costs are often recouped over the lifetime of the efficiency measures through energy savings, some consumers lack the financial means or the willingness to use their limited existing resources to make the initial purchase of high-efficiency measures.

These high first costs have been one impetus for utilities, states, and local governments to experiment with financing programs, where consumers are offered some form of program- supported financing to help pay for efficiency improvements.[5]

1. What is the rationale for offering energy efficiency financing?

(i.e. What problem(s) are you solving?)

Several cases may be made that today's EE financing market warrants tax payer or utility bill payer intervention due to either (a) market failure or (b) the broader set of goals (e.g., energy savings, emission reductions) of tax payer and bill payer funds relative to private monies, which are typically deployed purely to seek financial return. The problems that tax payer or bill payer-funded financing programs are seeking to solve need to be clearly identified and tested to ensure that (a) the problem(s) exists and (b) allocating funds to programmatic financing initiatives effectively addresses the problem(s).

Depending on a program's objectives and the willingness of private markets (e.g., lenders, investors) to respond to program initiatives, these interventions may be temporary or long-term. The motivations for these programmatic interventions are potentially numerous, but based on our experience reviewing financing programs, we discuss five common program rationales below:

A. New financial products are needed to overcome barriers specific to energy efficiency;
B. Some consumer market segments are under-served by private markets;
C. More information is needed before private markets can provide appropriate financial products;
D. Financial product standardization and aggregation are needed for the private markets to deliver attractive capital, and;
E. Larger consumer cost contributions are needed to increase the leverage of limited tax payer or utility bill payer funding.

Historically, Rationales A and B were typically the primary motivators for policymakers and program administrators to support EE financing initiatives as they sought to overcome barriers faced by individual consumers. Through time, these rationales have expanded to include broader "market failures" like those highlighted in Rationales C and D. More recently, the attention being paid to financing has focused on the need to increase consumer cost contributions to EE projects in order to achieve significant EE savings goals in the context of limited program budgets (Rationale E).

In the following sections, we describe each rationale and raise questions that program administrators should consider in program design and evaluation in order to determine whether program results support the rationale for operating them; and what, if any, program modifications are needed for the programs to effectively deliver on program administrator goals.

Rationale A: New Financial Products are Needed to Overcome Efficiency's Specific Barriers

The high up-front cost of some energy efficiency measures is one of several barriers to broader consumer adoption of these improvements. Some new financial products have special features with the potential to address both the high first cost barrier and other barriers such as renter/owner split incentives, long project paybacks, and balance sheet treatment that lead to consumer under-investment in EE in certain market segments.[6] For example, two novel financial products that have garnered significant attention are Property Assessed Clean Energy (PACE) and On-Bill Financing (OBF).[7,8] PACE involves financing energy improvements through a special property tax assessment, which is typically senior to all other debt on a property, including the first mortgage.[9] OBF involves repaying financing for energy improvements on the customer's utility bill, often secured by the possibility of service disconnection for non-payment.[10] These products' novel security may offer value to lenders and investors that can be leveraged to expand consumer access to attractive capital beyond that which private markets can deliver through traditional financial products.[11] In Table 1, we highlight other potential advantages of these novel financial products (in addition to their potential to broaden consumer access to capital)—and uncertainty about their value—relative to commonly-used traditional financial products.

Broadly-available energy performance guarantees and energy savings insurance products are also promising tools that can reduce the risk that a consumer will not realize the energy savings they are expecting. While guarantees and insurance products are not financial products themselves, they may help to catalyze the delivery of innovative financial products whose security is tied to these energy savings and energy service delivery models. For example, these products might enable third parties to finance the energy improvements and consumers to simply pay for energy savings as they are realized rather than taking on financing (and project performance risk) themselves.[13] Given their promise, there may be a policy justification for offering tax payer or utility bill payer funds to support the development and implementation of these energy savings performance risk reduction products in some markets.

Key questions:

- Are novel financing products more effective in overcoming barriers to energy efficiency adoption than traditional financing products?
- Do specific features such as novel security, threat of utilities disconnection, alternative underwriting, or others lead to lower consumer defaults and delinquencies or higher participation rates?
- Do energy savings performance risk reduction products lead to lower consumer defaults and delinquencies on financial products or higher participation rates?

Table 1. Comparison of PACE and OBF programs with standard existing financing products

Financial Product	Security[12]	Overcomes Renter/Owner Split Incentives?	Overcomes Long Project Paybacks That May Exceed Tenancy/Ownership?	Overcomes Balance Sheet Barriers?
Unsecured Loan	None	No	No	No
Mortgage	Lien on consumer's property	No	No	No
PACE	Super-senior lien on consumer's property	Maybe.* If lease contracts include the pass through of property taxes to tenants	Maybe.** PACE assessments are transferable from one property owner to the next.	Maybe. Uncertainty remains about whether PACE can be treated as an "off-balance sheet" obligation.
OBF	Tariff on property's (or unit's) utility meter	Maybe.* If tenants pay utility bills.	Maybe.** In some cases, OBF tariffs automatically or, with occupant consent may, transfer from one tenant or owner to the next.	Maybe. Uncertainty about whether OBF can be treated as an "off-balance sheet" obligation.

* The value of PACE and OBF for overcoming the "split incentives" barrier remains uncertain and is based on the assumption that tenants will value the installed improvements and be willing to pay for them through a charge on their utility bill or an increase in their rent.

** The value of "transferability" for overcoming the "long project payback" barrier remains uncertain and is based on the assumption that subsequent tenants/owners will value the improvements for which they are being asked to assume the obligation to make debt payments.

Rationale B: Some Consumer Market Segments Are Under-Served by Private Capital Markets

Many consumers have adequate access to attractive capital today to overcome efficiency's high first cost barrier.[14] However, others do not (e.g., small businesses, affordable multifamily property owners and tenants) (Bell et al. 2013). In many cases, private financial markets do not serve these consumers well or serve them only with relatively unattractive, high cost products because the perception is that lending to certain market segments represents too high a risk relative to the potential financial return. It may be that information, through the collection of performance data, will be sufficient to make financing more accessible to these consumers (see Rationale C below).

However, there are some market segments that may be deemed by private lenders as unprofitable to serve, regardless of better performance data. Even if EE financing outperforms relative to other types of financing, this outperformance may not be sufficiently large to fundamentally alter the costs and risks required to deliver capital to these more difficult-to-serve consumers.

While private monies typically seek purely financial return, tax payer and utility bill payer funds target a range of system and public benefits (e.g., cost effective energy savings, reduction of environmental impacts of electricity production, diversification of resource mix to reduce various risks). This more holistic view may lead to a different assessment of risk and return based on broader programmatic goals, and may warrant long-term provision of tax payer or bill payer direct loan capital, or credit enhancement to private markets, to deliver attractive capital to overcome barriers to adoption for hard to reach market segments.[15]

Key questions:

- What market segments are currently underserved by capital markets and why?
- Which market segments are likely to continue to be underserved even if the problems underlying other rationales are addressed?
- Can attractive capital be extended to underserved consumers at "acceptable" risk to those consumers and in a way that delivers low-cost energy savings to tax payers and utility bill payers?

Rationale C: More Information Is Needed for Private Financing Markets (e.g., Lenders, Investors) to Take over

A range of financial tools and capital providers already exist to enable consumers to borrow funds to pay for the up-front cost of energy efficiency improvements. However, the terms (e.g., interest rate, loan length) and underwriting criteria of these products may not reflect all of the potential benefits that financed improvements deliver to consumers. For example, energy savings from these investments reduce consumer utility bills – in some cases by more (over the life of the improvements) than the cost of the energy improvements themselves. This financial benefit, in theory, should reduce consumer defaults on financial products relative to financing for other types of activities (e.g., boat purchase, granite kitchen countertops) because it leaves consumers with more money with which to repay their debt.[16] Lower consumer defaults should yield some combination of reduced interest rates, longer loan lengths, and less restrictive underwriting criteria (so that more consumers qualify for financing). Lower interest rates and longer loan lengths would enhance project cash flows by reducing a consumer's regular interest and principal payments and might support broader consumer EE adoption and deeper per-project energy savings.

Today, however, financial institutions lack access to adequate data to assess and price both energy savings and the improvement in borrower financing repayment trends that these savings may deliver. EE financing programs have often been limited in scale, data recording methods have not been standardized and, since many programs were launched as part of ARRA, have not existed long enough to capture default rates over a full loan cycle (Hayes et al. 2011). This information asymmetry may lead to credit rationing (Palmer et al 2012), where private markets do not deliver an adequate supply of attractive capital to this market. Financial products whose terms and underwriting are based solely on consumer characteristics or the value of collateral (and not the potential energy saving benefits of the financed projects), may be relatively unattractive compared to those that would be offered if more information were available to financial institutions. These less attractive financial products may, in turn, inhibit consumer adoption of energy efficiency.

In this context, tax payer and utility bill payer-supported financing programs could be used as temporary interventions to deliver more attractive and accessible financial products than are available in private markets today, while developing the requisite data on both project energy savings and the impact of that energy savings on financing product performance. This data could be used to substantiate to financial institutions the benefits of offering financing for efficiency improvements and enable a transition to fully- private financing markets in the future that account for these attributes. It is important to note that EE financing programs have been operating for several decades and have not so far been structured or documented in a way that has led private capital providers to alter their risk assessments of this market (and, in some cases, program volumes have not been large enough to warrant their attention).

Key questions:

- Does financing for energy efficiency have lower consumer defaults and delinquencies than financing for other property improvements? If so, what is the cause of these differences (e.g., is it specific product characteristics, characteristics of early adopters, the presence of energy savings, or something else)?
- Is the performance of EE financing strong enough to warrant substantial improvements to the interest rates, loan lengths and/or underwriting for private financial products? Items unaffected by credit risk such as marketing, underwriting and back office processing often account for a substantial portion of financial product costs. If material financial product improvements are warranted, for which consumer segments or EE improvement types?
- What data are required to enable financial institutions to obtain sufficient evidence to improve the terms of their current product offerings? How long will it take to build this data set?[17,18]

Rationale D: Financial Product Standardization and Aggregation Are Needed for Private Markets to Deliver Attractive Capital

EE financial products, particularly those in the residential and small business sectors, tend to be quite small in terms of loan size. Financial institutions often participate profitably in markets like this by offering consumers standardized products that can be originated in high volume,[19] aggregated and re-sold to other investors through a highly-organized secondary markets transaction (which re-capitalizes financial institutions with sufficient monies to originate more loans or leases).[20] Today, however, the EE-specific financing market is characterized by low volume, lack of product standardization,[21] and the absence of vehicles to aggregate financing pools for re-sale.[22]

Tax payer and utility bill payer-supported financing programs could be used as a temporary or long-term intervention to standardize financial product terms across financial institution partners and/or to aggregate these financial products and facilitate secondary markets transactions. This access to secondary markets has the potential to deliver large pools of institutional investor capital for energy efficiency financing.

Key questions:

- What are the real barriers to the development of secondary markets for EE financing? A few efficiency programs have faced capital constraints due to high financing volume,[23] but most programs and their financial partners have substantial outstanding lending capacity.
- Will "self-organized" secondary markets pathways emerge without programmatic intervention if adequate consumer financing demand and product volume develops?
- Does sufficient consumer demand exist today to warrant program investments in aggregation and securitization infrastructure, or should interventions simply focus on increasing the volume of standardized loans?

Rationale E: Larger Consumer Cost Contributions Are Needed to Increase the Leverage of Limited Tax Payer or Utility Bill Payer Funding

The focus on financing by some policymakers and program administrators is driven primarily by a desire to encourage substantial cost contributions by participating consumers in order to stretch the impacts of limited tax payer and utility bill payer funding in the face of aggressive energy savings goals. Other financial incentives (e.g., rebates and tax credits) can also be effective in reducing efficiency's first cost hurdle and have proven their efficacy in driving consumer EE adoption. However, rebates deliver limited leverage and financing programs may increase this leverage.[24, 25] For example, programs offering financial institutions a 10 percent loan loss reserve have the potential to leverage each $1 of tax payer or bill payer funds into a total of $10 of investment in EE improvements (see Table 2).[26]

Generally speaking, however, the private market for financing property improvements is large, sophisticated and mature. "High first costs" may be an important barrier in some situations, but is there actually a market failure in delivering adequate pools of attractive capital through existing financial products that would provide the rationale for using tax payer and utility bill payer funds? Financing programs can only deliver on their leverage potential to the extent that they drive (or enable) consumer demand for EE. For many consumers and consumer classes, lack of **demand** for EE – not access to attractive capital to pay for these upgrades – may be the primary challenge. If program administrators reduce support for other program strategies in favor of financing, and consumer demand does not materialize, they risk missing their energy savings targets or other goals. Financing can (and often should) be combined with other strategies (labeling, rebates, contractor training, etc.), but the right mix of strategies is something that needs to be carefully considered and tested. Ultimately, the "consumer demand" issue is central to any strategy's potential to reach EE policy goals.

Key questions:

- Is financing an effective tool for driving consumer EE adoption? For which consumers and at what cost?
- What other strategies should be combined with financing to maximally increase demand at the lowest possible cost?

The rationales described in this chapter highlight the thought process and key questions that policymakers and program administrators should consider before launching new financing programs or committing to increasing their reliance on existing financing initiatives. It is also important to compare the realized cost and effectiveness of financing programs compared to other options for intervening in efficiency markets; we explore questions relevant to this level of inquiry in Chapter 3.

CHAPTER 3: IS FINANCING THE BEST (OR ONLY) OPTION?

Once rationale(s) for supporting an EE financing program are identified and evaluated, it is important to assess the overall effectiveness of financing in driving consumer adoption of EE relative to – or in addition to – other possible strategies.

The up-front cost of efficiency investments is just one of many barriers, and often times not the most important one.[28] A range of non-financing program strategies and other activities (e.g., rebates, technical assistance, labeling, codes & standards, workforce training, etc.) target other barriers to efficiency adoption such as lack of consumer understanding of EE benefits, uncertainty about energy improvement performance, or an inadequate supply of qualified EE service providers. As shown in Figure 3, financing is part of a holistic suite of strategies targeting multiple barriers to consumer EE adoption.

Table 2. Sample leverage potential of EE program funds allocated to rebates compared to credit enhancements

Program Incentive	Potential Leverage of Program Funds[27]
25% Rebate	4:1 (for every $1 rebate, $4 total is invested in EE)
50% Rebate	2:1
5% LLR	20:1
10% LLR	10:1

> **2. Does financing address key barriers better or at a lower cost than other options for intervention?**
>
> (i.e. Is financing the best option for solving this problem?)

It is important to recognize that developing and supporting EE financing program infrastructure can have substantial costs.[29] Thus, as a practical matter, program administrators (and policymakers) must often weigh and decide whether the decision to offer financial products will lead to budget reductions for other program strategies or elements, particularly if they are operating in a zero sum program budget environment.[30] Therefore, it is important that program administrators assess whether financing interventions can achieve program goals (e.g., scale, cost-effective energy savings, equitable consumer access to programs) as, or more, effectively – and at lower tax payer or utility bill payer cost – than these other strategies, and for which consumer segments.

Table 3. Key questions on the relative efficacy of financing in driving and enabling consumer adoption of energy efficiency

Question	Issue	Discussion
Are consumers as (or more) likely to adopt targeted EE improvements if offered financing rather than rebates (or other support such as technical assistance)?	There is little evidence today that financing is as effective (or more effective) in overcoming the fundamental barrier to EE (i.e., driving consumer demand) or that it can do so at lower cost than other program strategies (Borgeson et al. 2012).	From a consumer's perspective, rebates improve the economics of projects and have been demonstrated to drive EE adoption; financing, even with no interest, simply delays payment. From a program administrator's perspective, financing, if it leads to adequate EE adoption rates, *may* reduce program costs (and rate impacts) compared to rebate programs.
What impact does program-sponsored financing (rather than rebates or other incentives) have on the likelihood of consumers that already have access to capital to adopt targeted EE improvements?	Many households and businesses already have attractive private financing options at their disposal.	Does the availability of EE-specific financing drive these consumers to adopt energy efficiency? Relatively few participants utilize program-sponsored financing when they are required to choose between financing and rebates (Nadel 1990, Stern et al. 1985). If program administrators shift away from rebates towards financing, what impact will this have on over-all market penetration and participation rates? For those households and businesses that do not have access to attractive financing tools, will they be more likely to participate in EE programs if financing is offered?
Do projects deliver greater energy savings if program financing is used (or available) compared to rebates (or other strategies)?	Program administrators have multiple goals that often include both increasing the number of consumers adopting EE and increasing the depth of energy savings that each consumer is achieving.	Projects that deliver deep energy savings often have higher up-front costs. Attractive and accessible financing may be an important tool for driving those consumers that do adopt EE to make more comprehensive improvements; this hypothesis should be evaluated.

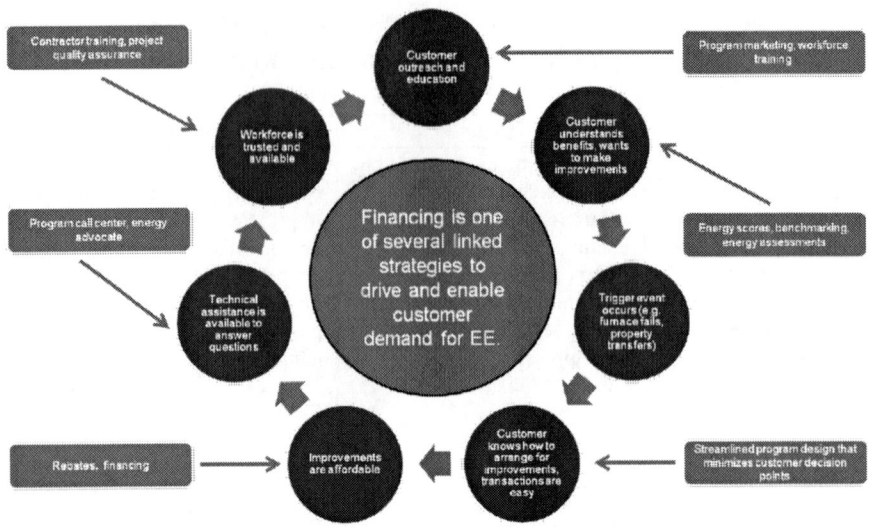

Figure 3. Strategies to drive and enable consumer demand for EE (SEE Action 2013).

In Table 3, we highlight several examples of financing program design questions and discuss their implications for the efficacy of financing relative to other program strategies in delivering energy efficiency. For illustrative purposes, Table 3 poses several either/or questions about the relative effectiveness of financing compared to rebates. We acknowledge that other strategies need not necessarily be abandoned in favor of financing and that a combination of strategies (e.g., rebates, financing) may be effective in driving consumer EE adoption while reducing overall tax payer or bill payer cost for these energy savings. A key challenge is developing the combinations of program strategies that can most effectively deliver low-cost energy savings in various customer market segments.

Assessing the relative value of financing compared to other interventions is not a simple activity; we discuss some methods for evaluating this important issue in Chapter 5. In most cases, program administrators will need to go beyond asking participants whether they "needed" financing to do a project or whether they "want" financing in addition to rebates or other support. It is easy to get the answers we want to these questions without necessarily obtaining an indication of the true efficacy of program- sponsored financing. Instead, program administrators will need to *test* different program offerings and observe who participates, who does not, and at what cost.

We also acknowledge that "temporal variability" (e.g., the value of supporting EE financing initiatives and the mix of financing products and programs one might choose to offer) may vary through time depending on evolving market conditions. For example, program-sponsored financing tools that rely on novel security (e.g., PACE, OBF) may be more effective during periods of weak real estate markets when households and businesses lack access to property-secured financing vehicles (e.g., mortgages, home equity lines of credit) that have traditionally supported much U.S. household and business borrowing for property improvements. Similarly, financing programs may be more effective during periods when private sector interest rates are high; thus, low-cost programmatic financial products are relatively more attractive compared to a market environment of low private market interest rates.

CHAPTER 4: WHAT FINANCING PROGRAM FEATURES BEST DRIVE DEMAND?

In addition to setting clear rationales for operating an EE financing program and evaluating whether financing is the most effective means for reaching program targets, new and existing financing programs can also benefit from better understanding which specific program design attributes are *most* effective in driving or enabling EE adoption– and for which consumers.

3. What specific financing program design features best drive demand for energy efficiency?
(i.e. How do you design the financing program for greatest impact?)

Table 4. Key questions exploring financing program design features

Question	Issue	Discussion
Do lower interest rates, longer financial product maturities and/or less restrictive underwriting than what is available in private markets *increase consumer adoption of* targeted EE improvements? How important is timely and streamlined loan approval to increasing consumer adoption?	These features may help drive consumer adoption of EE. However we lack basic information about consumer elasticity of demand around interest rates and loan terms,31 and little data has been collected about how relaxed (or alternative) underwriting criteria influence consumer likelihood of investing in EE. For some consumers and contractors, fast-close, easy-to-use financial products that can be closed at the consumer's "kitchen table" may be more effective in driving EE adoption than low-interest/long-term financial products with higher transaction costs (e.g., closing at the bank).	Substantial program administrator and policymaker attention and resources have focused on improving financial product access and terms (see Rationale C from Chapter 2). Better understanding of which product features are most important in increasing EE adoption—and for whom— would help to better target resources such as credit enhancements.
Is offering EE financing to consumers that lack access to capital in private markets more or less effective in catalyzing consumer adoption of targeted EE improvements than for other consumers?	Offering EE financing to consumers that lack access to other sources of capital to pay for these improvements may be more effective in driving consumer EE adoption than it is for the broader consumer base.	Consumers that lack access to sources of capital may be "debt averse" and more concerned about the consequences if energy savings do not materialize and they are unable to make debt service payments compared to other consumers. Significant resources are often allocated to expanding consumer access to capital, but, in many cases, the average program participant would qualify for existing private financial products. Better understanding of the non-financing barriers to EE adoption amongst consumers that lack access to attractive private financial products may enable limited

Question	Issue	Discussion
		program resources to be allocated more effectively.
Does the ability/ willingness to repay EE financing on a tax or utility bill increase consumer adoption of EE improvements relative to traditional financial products?	The consumer convenience of repaying financing on an existing tax or utility bill may reduce consumer debt aversion, facilitate the contractor sales process, or otherwise increase the uptake of EE improvements relative to offering financial products that are repaid on a separate bill. These benefits may be particularly effective with on-utility bill repayment, where a single bill might show the energy savings for which a consumer is making debt service payments and those payments.	Uncertainty remains about whether the primary value of novel financial products is their unique security features (which may improve product terms or expand access to capital) or their capacity to help drive consumer adoption of EE because they offer a more convenient repayment mechanism. Certain product features (e.g., primary residential PACE lien, on-bill tariff structure) may be politically or legally difficult to implement, and a better understanding of the value of these tools in increase consumer EE adoption would help policymakers understand the importance of various product features.
Does expected (or realized) "bill neutrality" increase consumer adoption of targeted EE improvements?	Expected bill neutrality (i.e.,– the expectation that consumer energy savings will be at least as large as consumer financing payments) may be a compelling tool for driving consumer adoption of EE, but its efficacy remains largely untested. Energy savings is often a sales "hook," but many consumers decide to move forward with energy improvements to solve other household or business problems (e.g., comfort, aging or failed equipment) (Fuller et al 2010). Realized energy savings also tend to vary from expectations and the consequences for market development of consumers	Requiring bill neutrality may restrict the depth of an energy savings project can achieve. Given uncertainty about their impacts on both driving consumer EE adoption and/or enhancing repayment trends, caution is warranted as this feature may "lock' programs out of delivering some of the deep energy improvements that may be necessary to achieve broad energy saving policy goals.

Table 4. (Continued)

Question	Issue	Discussion
	not realizing the expected level of energy savings are uncertain.	
Do energy performance guarantees increase consumer adoption of targeted EE improvements?	Energy performance guarantees that ensure a consumer will receive a specified level of energy savings (or insurance that covers energy savings underperformance) may be compelling tools for driving consumer adoption of EE, but their efficacy remains largely untested in most market sectors (with the exception of public/ institutional markets where ESCOs have been relatively successful).	It is estimated that Energy Service Companies (ESCos), which offer Energy Savings Performance Contracts (ESPCs) that guarantee consumers will achieve expected energy savings, have penetrated approximately 30 percent of local, state and federal EE markets and over 40 percent of K-12 schools (Larsen et al. 2013). ESCOs have had more limited success in offering ESPCs in other market sectors.
Is offering multiple financing products (or having multiple financial institutions offering the same product) more effective in driving consumer adoption of targeted EE improvements than offering a single financial product (or having a single financial institution partner)?	Some EE financing programs offer multiple financing products (or have multiple financial institutions offering similar products and competing for consumer business). Others offer a single financial product (or have a single financial partner). A single option might simplify the contractor sales process and avoid financing becoming an additional complex decision that consumers must make. However, some consumers may value the option to pick the financial product that best suits their needs from a suite of multiple program-supported product offerings. Having.	It remains unclear whether single or multiple options are more effective in driving consumer adoption of EE improvements. Reducing uncertainty about which approach is most effective, and for which consumers, would have substantial impact on program design as many administrators are pursuing "open market" models through which any qualified financial institution may compete to deliver EE financing products to consumers and/or contractors.

Question	Issue	Discussion
	multiple products or partners may also encourage competition and innovation	
Does automatic or optional transferability of financing payments increase consumer adoption of targeted EE improvements?	Transferability is the automatic or optional transfer of the obligation to pay a financing charge from one tenant to the next or from one property owner to the next. This feature may increase consumer willingness to invest in EE improvements with paybacks that exceed their expected tenancy in, or ownership of, a building because they may be positioned to transfer the remaining charge to the subsequent building occupant when they move.	If subsequent tenants & owners do not value the EE improvements, they may not accept the charge or may reduce the price they are willing to pay to purchase or occupy a property. This leads to uncertainty about whether transferability will increase consumer EE adoption. Substantial policymaker resources are often allocated to implementing transferable financial products despite the lack of evidence that consumers adopt EE or adopt deeper EE improvements when this feature is present.

This chapter identifies a number of key program design features (e.g., financial product interest rate, term, repayment mechanism, ease of use) and poses questions for administrators to consider about the extent to which these specific design features can **drive consumer adoption** of targeted efficiency improvements. These questions, along with some discussion on each, are included in Table 4.

Any feature that might affect consumer EE demand is important, but examining those that are most costly to program administrators to offer is a good place to start. It is important to know whether incentives such as interest rate buy downs or credit enhancements (e.g., loan loss reserves) actually have a significant positive impact in driving EE adoption. Other financing program elements are "free", but enabling their use requires expending political capital to pass legislation or change policy rules through regulatory proceedings (e.g., OBF, PACE); thus, it is important to better understand the impacts of these features on consumer demand.

In the next chapter we discuss some of the methods and resources for evaluating these questions.

CHAPTER 5. EVALUATING KEY QUESTIONS

Policymakers and program administrators in a number of states are interested in relying more heavily on EE financing, often as part of a strategy to increase the leverage of limited ratepayer or public monies. While financing programs are promising for some consumers, these initiatives have not been subjected to rigorous evaluation techniques. Before administrators make more substantial commitments to financing – particularly if those commitments require reducing investments in other program strategies – it is important that policymakers and administrators clearly define the specific problems their financing programs are designed to overcome and integrate evaluation techniques that can reduce the uncertainty about whether these initiatives are effective tools for driving consumer adoption EE at low long-term cost. Chapters 2 to 4 describe many of the key questions which must be explored to reduce the uncertainty about financing's effectiveness in delivering on policy maker and program administrator goals.

There are a range of techniques and levels of effort that can be applied to answering these questions. Some questions are best analyzed through qualitative approaches such as market research, discussion with program staff and stakeholders, and informal observation over time. However, to address other questions, more robust and quantitative approaches are necessary. These evaluation efforts require upfront planning, take time to assess, and may involve a significant $$ investment in program evaluation. In some cases they involve collecting and analyzing potentially sensitive consumer data or involve implementing controlled experiments. But, they are often the only way for us to know whether our interventions are working and can better enable policymakers and program administrators to make informed choices about how to allocate limited tax payer and utility bill payer resources.

In this chapter, we describe three broad categories of evaluation activities that program administrators and policymakers can consider. Evaluation categories are discussed in order of confidence that policymakers and administrators can have in drawing definitive conclusions from such activities:[32]

1. Qualitative market research (lower confidence);
2. Analysis of standardized financing program data; and
3. Experimental design and quasi-experimental design (higher confidence).

> ### *EXAMPLE:* SECONDARY MARKETS DEVELOPMENT
>
> In Chapter 2, we asked whether self-organized secondary markets might emerge in the absence of programmatic interventions if sufficient EE financing volume existed. In several states (e.g., New York, Pennsylvania), program administrators are utilizing public or quasi-public entities to aggregate EE financial products and ultimately facilitate their sale to secondary investors. Initial sales of these project loan pools to secondary buyers has required the provision of credit enhancements, which, depending on loan default rates may have substantial costs to program administrators. Whether these interventions are the best use of public or utility ratepayer resources is a question worthy of qualitative consideration by policymakers and administrators, but not one that lends itself well to rigorous testing. In cases with limited direct experience and significant uncertainty, one approach is to look to the past experiences of other emerging financial markets (e.g., time shares) for evidence on how EE finance markets might evolve and to make strategic decisions based on these other markets. Another approach is to implement "no regrets" actions, such as product and program standardization across jurisdictions, which are necessary precursors to the development of robust secondary markets, in order to buy time for private markets to respond on their own without the risk of over-committing programmatic resources to strategies that might ultimately prove unnecessary or ineffective. After a period of time, policymakers and program administrators could then revisit the question with additional data and experience in hand.

1. Qualitative Market Research

In some cases, policymakers and program administrators will have to make judgment calls on appropriate programmatic approaches using a range of qualitative techniques (e.g., research on experiences from other emerging financial markets, interviews with program participants and potential financial partners, and focus groups). Qualitative market research can be an effective way to understand how potential market participants and consumers think about EE financing and their perspectives on the importance of various program elements. The text box below provides an example of a key question—the need for programs to support the development of secondary markets—for which qualitative research might be the best approach to resolving uncertainty.

2. Analysis of Standardized Financing Program Data

The standardization of financing program data collection and analysis is an important approach to resolving several foundational questions regarding the performance of financing for efficiency projects. This approach is best suited to answering broad questions whose answers are unlikely to vary dramatically across small differences in specific program design elements or financial product features. In the text box below, we describe how standardizing the collection of EE project and loan data can be used to better understand the performance of EE financing. There is currently active research and inquiry by the U.S. Department of Energy and the State of California into what financing data can and should be collected to enable better analysis of EE financing programs. The standardization of data collection may have co-benefits, such as informing efforts to reflect the value of EE improvements in property values.

EXAMPLE: EE FINANCING DEFAULT RATES

In Chapter 2, we noted that, at present, some/many financial institutions have claimed that they lack access to adequate data on actual vs. estimated energy savings from efficiency projects and the improvement in borrower financing repayment trends that these savings may deliver. Collecting and analyzing data from the increasing number of EE financing programs operating in local and regional markets may be able to reduce this information gap. However, no single program is large enough (in volume or geographic coverage) to deliver the large data sets across geographies that are necessary for financial markets to assess whether consumers default or become delinquent at lower rates for EE financing than for other financial products. The diversity of program sponsors and lack of clear protocols for collecting and sharing data across programs makes assembling this EE financing performance data challenging.

Standardizing data collection and analysis protocols across these financing programs is a powerful tool for aggregating sufficiently large pools of data to bridge this information gap.

> This standardization will ensure that the performance data from pools of EE financing in New York is relatively similar to that from pools of EE financing in California – and that this data is broadly available for analysis. In the event that analysis of data from these financing programs does not demonstrate conclusively that efficiency financing programs out-perform other financial products to warrant better product terms from financial markets, then program sponsors may need to use experimental design techniques to assess what specific financing program, project and consumer attributes are most likely to deliver on EE financing's promise of low default rates (see next section).

3. Experimental and Quasi-experimental Design

Across the country, most EE programs deliver a range of services (e.g., rebates, financing, technical assistance, contractor training) that may impact the consumer's propensity to invest in – or repay financing for – EE improvements. The higher the number of factors that may influence target consumer behaviors (e.g., EE adoption, financing repayment), the more difficult it is to identify the impacts of *specific* program design features on the desired program outcome.

Qualitative assessments are unlikely to yield answers to many of the program design questions described in this report with sufficient confidence that it would be prudent for administrators and policymakers to rely solely on them in designing a finance program or making substantial shifts in resource allocations towards financing. Similarly, standardizing data collection and analysis is unlikely to yield answers to research questions related to the efficacy of specific program design elements or financial product features in driving consumer EE adoption. Instead, experimental and quasi-experimental design techniques are the best techniques for answering the detailed questions identified in Chapters 3 and 4 about whether EE financing can drive consumer adoption of cost-effective EE at lowest public or ratepayer cost, and what specific financing program features are most effective in achieving administrator goals. As policymakers and administrators increasingly seek to assess the cost-effectiveness of EE financing programs, resolving these foundational questions is essential.

Experimental and quasi-experimental design approaches hold as many program delivery factors as possible constant in order to isolate and study the impact of specific financing program features. Both experimental and quasi-experimental designs typically create two groups of consumers, one of which is given the treatment[33] (the treatment group) and another which is not (the control group). The key difference between the two approaches is the method used to assign participants to the treatment and control groups. **Experimental design, or a randomized control trial (RCT), uses random assignment** of participants to the two groups while **quasi-experimental design uses non-random assignment** of participants to the two groups.[34] Where program administrators expect one program design or financial product feature to be significantly more effective than another (i.e., a large effect size), quasi-experimental design may be preferable because it often avoids some of the implementation challenges that a RCT poses for administrators (e.g., market confusion, equity concerns across consumers or administrative hassle).

However, there is substantial uncertainty about the answers to many of the questions raised in this report, and program administrators should consider experimental design (i.e., RCT) both because it can detect smaller effect sizes and because it delivers the highest level of confidence that the results of the experiment are representative of the efficacy of the program design element in question at larger scale. In the text box below we provide an example of how experimental design can be used to explore the impact of bill neutrality on consumer demand. We also provide an introduction to experimental design in this section and more detailed information in Appendix A (Experimental Design Methods & Practical Experimental Design Implementation Guidance).

EXAMPLE: BILL NEUTRALITY

Bill neutrality (i.e., requiring that a project's expected energy savings exceed principal and interest repayment costs) may be an important driver of both consumer adoption of energy efficiency and strong financial product performance (i.e., low default rates). However, bill neutrality can restrict administrators' abilities to achieve their energy savings targets by limiting the depth of energy savings that consumers can pursue, and it may prevent consumers from investing in energy efficiency improvements that offer other perceived benefits (e.g., comfort, aesthetics, health and safety). In addition, expected bill neutrality's impact on financial product performance is uncertain given variance in realized savings from expectations and other consumer credit considerations that may play a larger role in influencing financing repayment trends (e.g., job status).

Experimental design is a powerful tool for resolving uncertainty about how bill neutrality both impacts consumer adoption of energy efficiency and consumer repayment of energy efficiency financing. Consumers could be randomly assigned to two groups, one of which is offered a package of improvements that is expected to be bill neutral and another of which is not. With random consumer assignment, any difference in outcome between the two groups can be attributed to relative efficacy of the two offers.35 While answering the question about financial product performance will take several years, program administrators will learn about whether expected bill neutrality is an important driver of consumer EE adoption.

Introduction to Experimental Design

The goal of experimentation is to see how well specific program designs work relative to other program designs. Good experimental design creates a way to test which program or program element more effectively delivers on policy maker and program administrator goals. Experimental design is often called A/B testing: comparing the outcomes from Offer A to Offer B (e.g., a financing program vs. a rebate program, a financing program with a loan interest rate of 10% vs. 5%, etc).

Experimental Design Basics
In an ideal hypothetical world, we would offer consumers one type of program or incentive and observe their response, and then go back in time and observe how those same consumers would respond to a different type of program or incentive (this other program type or incentive is often called the *counterfactual*). Comparing the outcomes from the two different programs offerings would tell you which program or incentive design is more effective. Unfortunately, we cannot observe this counterfactual.[36] This leaves us to use experimental design to compare two different (but similar) groups of consumers: one group of consumers given Offer A, and another group given Offer B. If the consumers are sufficiently similar, the difference in outcomes between the groups is a good estimate of the offers' relative efficacy.

The Selection Bias Problem

Why is experimental design necessary at all -- for example, could a program administrator assess the efficacy of the two offers by just giving everyone the choice of rebates or financing and see which delivers more projects or deeper per-project energy savings?

Table 5. Design techniques for selecting control and treatment groups

Design	Type of design	Confidence[39]	Description
Random Assignment of Consumers (Randomized Control Trial - RCT)[40]	Experimental	High confidence in evaluation of any key question	Consumers are randomly assigned to receive Offer A or B. Neither the contractor nor consumer is aware of which offer the consumer will receive before developing a potential work scope. This method will deliver the highest degree of confidence that the results of one's experiment are valid.
Random Assignment of Contractors	Experimental	High confidence in evaluation of any key question	Contractors are randomly assigned to deliver Offer A or Offer B to their consumers. This method will also deliver a high degree of confidence that the results of one's experiment are valid, but requires a large number of contractors.
Cutoff Point[41]	Quasi-Experimental	Relatively high confidence in evaluation of any key question	Consumers are assigned to receive Offer A if they are above a pre-determined cutoff point or Offer B if they are below a pre-determined cutoff point. The cutoff can be any continuous variable common to all potential participants (e.g., whether the second letter of their last name is before or after M).
Geographic Location[42]	Quasi-Experimental	Relatively low confidence in evaluation; confidence depends on similarities in geographic locations	Consumers in one geographic location are given Offer A, and those in another geographic location are given Offer B. The more similar the locations (e.g., demographics, climate, etc), the higher confidence the results.
Time Differences	Quasi-Experimental	Low confidence for evaluation of many key questions	Before a certain pre-determined date, consumers are given Offer A, and after the date they are given Offer B.

In most cases, the answer is a *no*, due to *selection bias;* the risk that consumers that choose Offer A may have fundamental differences from consumers that choose Offer B.[37] For example, the type of households that opt for rebates (Offer A) rather than financing (Offer B) may be fundamentally different than the type of households that are more likely to choose Offer B – let's say they tend to be more interested in getting deeper energy savings. If this is the case, we might assume that rebates result in more energy saving investments than financing, even though this is not the case – it is simply that those households would have made more energy saving investments regardless of the program offer. Without experimental design, comparing the outcomes between self-selecting groups delivers results that are biased and that are unlikely to reflect the efficacy of Offers A versus B. That is, the differences in the perceived effectiveness of the two offers could be largely due to differences in the participating consumers, not the offers themselves.

Using Experimental Design to Overcome Selection Bias

There are various experimental and quasi-experimental designs that attempt to deal with selection bias (see Table 4). All of these designs are fundamentally targeted at creating two groups of similar consumers to compare to each other. Experimental design eliminates the selection bias issue completely by randomly creating two similar groups (leading to high confidence in the validity of the results); quasi-experimental design uses non-random group selection techniques that will typically lead to results with lower confidence in their validity.[38]

A range of considerations will impact which, if any, of these methods is most appropriate:

- **Desired level of confidence in validity of results:** Some methods allow evaluators to have a high degree of confidence that the evaluation is valid; other methods are only valid for specific types of research questions. The level of confidence depends on how similar the treatment and control groups are to each other. With an RCT, the treatment and control groups are exactly the same (in expectation), while for the Geographic Location method, the control and treatment groups are from different locations and may be very different.

- **Ease of implementation:** Some methods may fit in easily with a standard program implementation method, while others may require a very specific implementation method. RCTs require specific, randomly assigned consumers to receive different type of program offers; these consumers must be tracked. In contrast, the Time Differences method simply requires consumers to receive one type of offer before a certain date, and another type after.
- **Ease of data analysis:** Random assignment methods are transparent and straightforward to analyze; other methods require more difficult analyses in order to attempt to correct for inherent differences between the control and treatment groups. For example, with a Geographic design, the analysis must control for demographic and energy use characteristics of the control and treatment groups; Time Differences requires controlling for all external factors that may have occurred. The Cutoff Point requires a regression discontinuity analysis.
- **Number of consumers required:** Each research question and experimental design requires a specific number of consumers ("sample size") in order to get results that are statistically significant.[43]

Choosing an appropriate technique for selecting the control and treatment groups is one of seven steps to integrating experimental design into EE financing programs (see Appendix A for more detailed discussion on experimental design).

CHAPTER 6. CONCLUSION

It is important for administrators to challenge and verify their assumptions before making fundamental shifts in their program offerings and then consistently evaluate whether financing, and what financing program designs, are most effective in moving consumers to action in implementing energy efficiency upgrades through time. This report offers a starting place for developing a better understanding of financing's role in driving cost-effective energy efficiency adoption. We encourage program administrators and policymakers to identify those issues and questions that are most relevant to their program's success and to begin to test whether their assumptions are correct. Not every program needs to answer every question – as more and more programs actively explore these questions, lessons learned can be shared. The answers to some key questions may not vary dramatically between programs, and administrators should consider coordinating their financing evaluation efforts regionally or nationally (through new or existing forums) to take advantage of likely economies of scale. However the answers to some of these questions may differ across – and sometimes within – both consumer classes and geographies. Thus, the conclusions from program evaluation should not be overly generalized.

APPENDIX A. EXPERIMENTAL DESIGN METHODS & PRACTICAL IMPLEMENTATION GUIDANCE

Appendix A describes the five experimental design methods outlined in Chapter 5 in greater detail and provides a seven step guide for integrating experimental design into EE financing programs.

The experimental design methods differ in their techniques for assigning consumers to the control or treatment groups (see Table A-1).

Random Assignment of Consumers

The gold standard of experimental design, and the most rigorous way to limit selection bias, is to *randomly assign* consumers to receive either Offer A or Offer B. With random consumer assignment, any difference in outcome between the two groups can be attributed to relative efficacy of the two offers.[45] For this type of experimental design, consumers are randomly assigned to receive one of two offers, A or B (e.g. rebates or financing; 5% or 7% interest rate loans).

Choosing when to conduct randomization and when to present the program offer to a customer is a function of the question being tested and ease of implementation for any given program. For example, a program administrator may wish to test how a consumer responds to different types of offers that a contractor makes (e.g., what causes consumers to complete more upgrades: an offer of financing or rebates). In this case, the key aspects of the experimental design are:

Table A-1. Design techniques for selecting control and treatment groups

Design	Type of design	Confidence[44]	Description
Random Assignment of Consumers (Randomized Control Trial - RCT)	Experimental	**High confidence** in evaluation of any key question	Consumers are randomly assigned to receive Offer A or B. Neither the contractor nor consumer is aware of which offer the consumer will receive before developing a potential work scope. This method will deliver the highest degree of confidence that the results of one's experiment are valid.
Random Assignment of Contractors	Experimental	**High confidence** in evaluation of any key question	Contractors are randomly assigned to deliver Offer A or Offer B to all of their consumers. This method will also deliver a high degree of confidence that the results of one's experiment are valid, but requires a large number of contractors.
Cutoff Point	Quasi-Experimental	**Relatively high confidence** in evaluation of any key question	Consumers are assigned to receive Offer A if they are above a pre-determined cutoff point, or Offer B if they are below a pre-determined cutoff point. The cutoff can be any continuous variable common to all potential participants (e.g., whether the second letter of their last name is before or after M).
Geographic Location	Quasi-Experimental	**Relatively low confidence** in evaluation; confidence depends on similarities in geographic locations	Consumers in one geographic location are given Offer A and those in another geographic location are given Offer B. The more similar the locations (e.g., demographics, climate, etc), the higher confidence in the results.
Time Differences	Quasi-Experimental	**Low confidence** for evaluation of many key questions	Before a certain pre-determined date, consumers are given Offer A and after the date they are given Offer B.

- Consumers are randomly assigned to receive either Offer A or Offer B;
- Consumers are not initially aware of which offer they will receive; they learn this only during the contractor's "pitch"; and
- Contractors do not choose which households to approach based on which group they are in and therefore which offer they will receive.[46]

One common technique for implementing random assignment is to set up an easy-to-access system that allows contractors to determine whether a consumer should receive Offer A or Offer B at the time the contractor is making the offer (e.g., contractors could call a 1-800 number, use an iPhone app). The consumers could be pre-randomized at the start of the experiment, with these assignments maintained in a database, and only made available to a contractor once a consumer has been engaged. Alternatively, consumers could be randomized at the moment the contractor is about to make the pitch and recorded in a database at that time.

If an administrator wants to test what offer motivates consumers to show more interest in energy efficiency (i.e., is advertising rebates or financing more likely to cause consumers to call a contractor to setup an energy assessment?), the administrator would pre-randomize consumers and then target the control or treatment group offer to consumers up-front, perhaps through a mailed advertisement.

Potential Challenges

The analysis method used to assess the efficacy of the two offers is relatively straightforward and transparent and involves comparing the average results of the group that received Offer A to the group that got Offer B.[47] The main drawback to this method is that it may be practically difficult to implement as contractors may be reluctant to add uncertainty to their sales process. Consumers may also hear (and complain) that other consumers were offered a different deal. Program administrators may want to consider testing Offers whose net economics are similar (e.g. $1,000 rebate v. $1,000 interest rate buy down) to reduce concerns about fairness. In cases where program administrators are concerned that consumers or contractors will be unwilling to accept randomization, using a Cutoff Point design is often the second best approach.

Cutoff Point

A good alternative to randomized assignment is assigning consumers to the control or treatment group based on a **cutoff point** – consumers above the cutoff point get Offer A and consumers below the cutoff point get Offer B. This works because consumers very close to the cutoff on either side are likely to be similar to one another. The cutoff point can be based on anything that takes on a continuous range of values and that the contractor is not aware of before they approach a consumer. The best cutoff points are those that are likely to have very little relationship to consumer characteristics that might be connected to the research question (e.g., is the third letter of the consumer's name before or after the letter "M" rather than does the consumer earn more or less than $50,000) because it increases the likelihood that consumers far from the cutoff point will resemble those that are closer to it. This method relies primarily on comparing the outcomes of those consumers close to the cutoff point on either side.[48] While we can have a high degree of confidence about the results from this method for the consumers close to the cutoff point, in order to extend the results to all consumers, we must assume that consumers that are far away from the cutoff point will react similarly to those near the cutoff point.

Potential Challenges

This type of experimental design is a very good alternative when randomization is not feasible. It will ensure a relatively high degree of confidence that the results are valid and is easier to implement. It is important that contractors are unaware of which side of the cutoff the consumer is on before beginning the energy efficiency sales process. If contractors are aware of which side each household is on, they might only sell to those consumers that have been selected to receive Offer A, biasing the results. Ideally, contractors would ask consumers for information at the point of sale that would allow them to determine which program offer to make.

Geographic Location

A third experimental design choice is assigning consumers to Offer A or Offer B based on geographic location (e.g., consumers in one neighborhood in a utility territory versus consumers in another similar neighborhood in a utility territory). Results from this experimental design are only valid insofar as one believes that the consumers in the different locations are similar. For example, if the geographic dividing line is in the middle of a street, so that one group is across the street from the other, consumers may be very similar. However if the geographic dividing line is state or county lines, those lines may be drawn in places with distinctly different demographics (e.g., dividing urban and rural locations, or wealthy and poor communities). The consumers in two different counties may be very different and the results from comparing the two groups to each other may therefore not be valid. Another issue is that there is no way to hide consumer locations from the contractors. Contractors will therefore be aware of which consumers will receive which Offer ahead of time, which risks biasing the results because differences in outcomes could be driven by contractors not consumers.

Potential Challenges

This type of experimental design may be acceptable if two locations with similar consumers can be found and could be preferable compared to not doing a pilot program that is evaluated. Because there is little that can feasibly be controlled in this design, implementation is relatively easy. The type of analysis for this design should attempt to control for every observable consumer trait (e.g., income, historical energy use, current energy use, home sale price, age, number of occupants, etc.) that may vary across geographic locations, and is therefore somewhat difficult.

Time Differences

A fourth choice is assigning consumers to Offer A or B based on time differences. Those that apply before a certain date are assigned to Offer B and those that apply after that date are assigned to Offer A. This type of design should only be used in the case that Offer A is known with certainty to be better than Offer B, but program administrators are seeking to better understand the magnitude of the difference on consumer EE adoption patterns. For example: Offer A is 5% financing, Offer B is 10% financing; 10% is offered before a certain date, and 5% is offered after that date. Consumers must be unaware that the offer will switch from Offer A to Offer B on a certain date. If possible, the date that the offer switches from A to B should be during a time that a change in outcomes is most attributable to the change in programs. For example, if most building retrofits occur during the summer, then the date should be in the middle of the summer so that changes in consumer activity can be observed apart from seasonal changes in demand.

Potential Challenges

This type of experimental design will only produce results that are meaningful for one very specific type of research question: if A is known to be better than B, and B is offered first and A is offered second. It will produce invalid results if the research question is whether A or B is better, or if A is known to be better and A is offered first. Even for the specific research questions for which this test is appropriate, the experimental design is still somewhat problematic and there is likely to be uncertainty about the validity of the experiment's results. This type of analysis attempts to control for every observable consumer trait (e.g., income, historical energy use, current energy use, home sale price, age, number of occupants, etc.), and every observable event that changed over time (e.g., employment rates, interest rates, etc.) and is therefore challenging.

This design has several issues. If consumers are aware that there is going to be a different program available after a certain date, consumers may postpone or accelerate energy upgrades to correspond with the program offer that they prefer. This phenomenon could lead to a large selection bias because the consumer groups would not comparable. Even if consumers are unaware that the program will change from B to A at a certain date, assigning consumers based on a cutoff date is challenging because some energy improvements may be investments that consumers make infrequently – perhaps every 10 or 15 years or longer. This is not a decision that a consumer makes every day or month.[49] Moreover, in practice, it may not be easy to offer the "better" program second; consumers who decided to get retrofits with the first offer may be unhappy once they realize that a better program is now offered (e.g., lower financing interest rates).

Randomize at Contractor Level

Another option is to randomly assign contractors to different programs, rather than assigning consumers to different programs. For example, contractors randomly assigned to group A are allowed to offer consumers rebates, while contractors in group B are allowed to offer consumers financing. This type of design answers slightly different questions than the previous designs. Rather than answering whether providing *consumers* with Offer A or Offer B is better, it answers whether giving *contractors* the ability to sell Offer A or Offer B is better. Because this design randomizes contractors, there will be a high degree of confidence in the validity of results.

Potential Challenges

The main issue with this type of design is that because the randomization is done at the contractor level rather than the consumer level, there must be many more participating contractors than in other designs.[50] Randomly assigning contractors is relatively easy to implement, if a large enough group of contractors can be found. Because it is randomized, the results will be valid and robust. The analysis must account for the fact that consumers targeted by one contractor may be different than consumers targeted by another contractor and so it is a slightly more difficult analysis than a design that randomizes at the consumer level.

Practical Experimental Design Implementation Guidance

This section offers a seven step guide for integrating experimental design into EE financing programs (see Figure A-1).[51]

Figure A-1. Overview of seven steps for effectively integrating experimental design into EE financing programs.

1. Define a Clear Research Question

The first step in integrating experimental design into an energy efficiency financing program is deciding on a specific research question to test. It is important to make sure that the research question is clear and precise. There are three keys to developing a strong research question:

- Define what you are comparing;
- Define the outcome to be measured; and
- Define the consumers that you are targeting.

What are you comparing?

Energy efficiency programs do not exist in a vacuum; their effectiveness is always relative to other types of initiatives or no initiative at all. For example, a new financing program could be compared to a rebate program, compared to no program (i.e. no financing or rebates), or compared to the status quo program (e.g., on-bill financing v. unsecured off-bill loans).

- **Weak Research Question**: "Is on-bill financing effective?"
- **Strong Research Question**: "Does the availability of on-bill financing increase consumer EE adoption rates in the single family residential sector relative to the availability of unsecured financing?"

What outcome will you measure to determine which program offer is more effective?

It is important to be clear about what outcomes will be measured to determine which program offer is more effective. For example, you might care about the performance of a pool of unsecured loans versus a pool of on-bill financing loans. In this case, the outcome that you would want to measure is default rates or number of late payments. Another outcome might be which program results in more contractor conversions of consumer leads to consumer projects. You might also want to measure the types of projects that are being completed through each offer and the level of energy savings they are achieving.

- **Weak Research Question:** "Is on-bill financing effective relative to off-bill unsecured financing?"

Table A-2. Sample outcomes to be measured in evaluating the effectiveness of financing programs

Outcome Measure	Data Collection Needed
Which offer results in more consumers adopting EE improvements?**	Whether or not each consumer presented with an offer adopted EE improvements
Which offer results in deeper per-project EE savings?**	Depth of savings for each consumer that adopted EE improvements; analysis of utility bills
Which offer results in specific targeted types of retrofits (e.g., "deep" retrofits vs. lower cost retrofits with fewer measures)	Whether or not each consumer completed specific types of home upgrade (e.g., whether each consumer chose to perform weather-stripping, wall insulation, window replacements, etc.)
Which offer results in lower-default rates on financial products?	Whether or not each consumer defaulted on loan payments over the life of the financial product
Which offer results in more cost effective, energy savings?	Per-project program expenditure relative to per-project energy savings

** This data should be collected for any research question.

- **Strong Research Question:** "Does the availability of on-bill financing result in higher per-project energy savings relative to standard financing in the single-family residential sector?"

Explicitly stating the outcome of interest will help to make clear what data needs to be collected. Data accessibility should, therefore, be an important consideration when defining the outcome measure. Table A-2 includes examples of outcome variables that may be of interest to program administrators.

Which consumer classes are you targeting?

The answers to many of the questions raised in this report are likely to vary across (and sometimes within) consumer classes. It is important to appropriately target experiments to ensure that the results are not overly generalized.

For example:

- The way in which a large institutional consumer responds to a five versus seven percent interest loan is likely to differ markedly from the way in which a single-family homeowner responds to these interest rate differences.
- Middle income households may be more motivated to pursue energy upgrades (or invest in deeper upgrades) by a program financing offer

than a rebate incentive relative to their higher income peers (who may be more likely to have ready access to attractive capital).
- "Bill neutrality" expectations or guarantees may increase a high energy use household's willingness to take on financing for deep energy saving projects compared to these features' impacts on low energy users.

Defining which consumer classes you are targeting is an important part of a strong research question.

- **Weak Research Question**: "Does on-bill financing result in more consumer spending on energy efficient products relative to standard financing?"
- **Strong Research Question**: "Does on-bill financing result in more spending by single family residential consumers on energy efficient products relative to standard financing? Does this differ for high income households relative to middle & low income households?"

Table A-3 includes examples of different consumer market segments within the single family residential sector; similar breakdowns could be done for multifamily and non-residential consumers.

2. Select an Experimental Design

The goal of experimental design is to answer the research question that you have developed. There are several tradeoffs to consider when deciding on the appropriate experimental design: which design produces results that are more robust and valid, which design is easiest to implement, which design requires the easiest analysis, and which design requires a smaller number of consumers.

3. Plan for Sample Size

Each research question and experimental design requires a specific number of consumers ("sample size") in order to obtain results that are statistically significant. It is essential that the sample size is planned in advance;[53] if there are too few consumers, then effort may be wasted on designing and implementing an experiment because too little data will be collected to complete an analysis. The easiest way to ensure that there will be enough consumers is to structure the pilot so that it operates for an open-ended amount of time until that sample size has been reached.

4. Plan for Data Collection

In addition to obtaining a sufficient sample size, the appropriate data must be collected in order to answer each research question. It is critical that program implementers develop and vet a data collection plan up-front to ensure that their experimental design efforts will yield results.

If the research question is stated in a clear and precise way, the data needed to answer the question should be evident. For **any research question**, two pieces of data are necessary at a minimum:

(1) whether the consumer received Offer A or Offer B; and
(2) whether or not the consumer accepted the offer.

Table A-3. Examples of single family residential consumer market segment

Consumer Grouping	Data Collection Needed
High income vs. low income consumers	The income band that each consumer belongs to (e.g., less that 50k, 50-75k, etc.) or the census block of each consumer that can be linked to geographic income data from the U.S. Census[52]
Elderly vs. young consumers	The age band for every consumer (e.g., less that 18, 18-25, 25-35, etc.), or the census block of each consumer that can be linked to age data for the head of household from the U.S. Census
Owners vs. renters	Whether each consumer is a property renter or owner or the census block of each consumer that can be linked to ownership data from the U.S. Census
High vs. low floor area	The square footage of each home
Other residential consumer characteristics	Data for each consumer on characteristics of interest that might play a role in their program participation, or U.S. Census block data for each consumer on: • Employment status • Educational status • Number of occupants • Value of home • Existing heating or cooling system

Depending on the **outcome measure** defined in the research question, additional data will be needed (see Table A-2 for examples of data needed for specific outcome measures). Depending on the **target consumer classes** defined in the research question, additional consumer characteristic information may be needed (see Table A-3 for examples of relevant consumer information that may need to be collected). It is very important that all necessary data is collected for every consumer that is offered Offer A or Offer B, *even if* they don't accept the offer, and even if they don't decide to do any energy improvements.

There are various ways to collect the necessary data; often data may be available through existing program processes and protocols. Contractors, utilities or financial institutions are often responsible for ensuring that accurate data is submitted in a timely fashion. Program administrators should consider whether a small financial incentive is appropriate to encourage contractors (or other third parties) to provide this data (and to compensate them for the added time it might take to submit it).

5. Implement

Consumers should be assigned to a control or treatment group and care should be taken to ensure that consumers receive the appropriate program offer. It is useful to keep track of all steps and procedures that are followed during the experiment to refer to during the evaluation.

6. Evaluate

Depending on the type of experimental design selected, different types of analyses are required. An experienced evaluator will be able to offer advice on the type of analysis required for the specific design and the level of experience and technical competence required.

7. Use Findings to Alter Program Design (or Setup New Experiment)

If properly designed, evaluations can provide administrators with useful information on the role of financing (or specific financial product features).

Administrators can use these results to alter program design or to setup additional experiments to hone in on program offers that will be most effective in driving different consumers to efficiency at lowest ratepayer or public cost.

REFERENCES

Bell, C., Sienkowski, S. & Kwatra, S. (2013). *"Financing for Multi-Tenant Building Efficiency*: Why This Market is Underserved and What Can Be Done to Reach It"*. American Council for an Energy-Efficient Economy (ACEEE).

Borgeson, M., Zimring, M. & Goldman, C. (2012). *"The Limits of Financing for Energy Efficiency."* Lawrence Berkeley National Laboratory.

Borgeson, M. & Zimring, M. (2013). *"Financing Energy Upgrades for K-12 School Districts"*. Lawrence Berkeley National Laboratory. LBNL-6133E.

Cadmus Group. (2012). *"California 2010-2012 On-Bill Financing Process Evaluation and Market Assessment."* Prepared for California Public Utilities Commission.

California Public Utilities Commission (CPUC). (2013). *"Decision Implementing 2013-2014 Energy Efficiency Financing Pilot Programs."* Decision 13-09-044 Issued 9/19/2013.

Connecticut Clean Energy Finance and Investment Authority (CEFIA). (2013). *"Comprehensive Plan*: FY 2013 through FY 2015."

California Public Utilities Commission (CPUC). (2013). *"Proposed Decision Implementing 2013-2014 Energy Efficiency Financing Pilot Programs."*

Cuomo, M. (2013). *"NY Rising, 2013 State of the State."*

Eto, J. & Golove, W. (1996). *"Market Barriers to Energy Efficiency: A Critical Reappraisal of the Rationale for Public Policies to Promote Energy Efficiency.* Lawrence Berkeley National Laboratory. LBNL-38059

Fuller, M. (2009). *"Enabling Investments in Energy Efficiency: A study of energy efficiency programs that reduce first-cost barriers in the residential sector."* Prepared for California Institute for Energy and Environment and Efficiency Vermont.

Fuller, M., Kunkel, C., Zimring, M., Hoffman, I., Soroye, K. & Goldman, C. *"Driving Demand for Home Energy Improvements."* Lawrence Berkeley National Laboartory LBNL-3960E.

Hayes, S., Nadel, S., Granda, C. & Hottel, K. (2011). *"What Have We Learned from Energy Efficiency Financing Programs?"* American Council for an Energy Efficient Economy (ACEEE).

International Energy Agency. (2008). "*Promoting Energy Efficiency Investments: Case studies in the residential sector.*" ISBN 978-92-64-04214-8.

Jaffe, A. & Stavins, R. (1994). "The Energy Efficiency Gap: What does it mean?" *Energy Policy, 22* (10), 804-810.

Nadel, S. (1990). "*Lessons Learned: A Review of Utility Experience with Conservation and Loan Management Programs for Commercial and Industrial Consumers.*" American Council for an Energy Efficient Economy (ACEEE).

Palmer, K., Walls, M. & Gerarden, T. "*Borrowing to Save Energy: An Assessment of Energy-Efficiency Financing Programs.*" Resources for the Future.

State and Local Energy Efficiency Action Network Financing Solutions Working Group (SEE Action). (2013). "Using Financing to Scale up Energy Efficiency: Work Plan Recommendations for the SEE Action Financing Solutions Working Group." Prepared by Lawrence Berkeley National Laboratory and Harcourt Brown and Carey.

Stern, Paul C., Elliot Aronson, John M. Darley, Daniel H. Hill, Eric Hirst, Willett Kempton & Thomas J. (1985). Wilbanks, "The Effectiveness of Incentives for Residential Energy Conservation," *Evaluation Review*, (April 1985, Volume 10, Number 2).

Stuart, E., Larsen, P., Goldman, C. & Gilligan, D. (2013). "Current Size and Remaining Market Potential of the U.S. Energy Service Company Industry." Lawrence Berkeley National Laboratory. LBNL-6300E, Harcourt Brown & Carey, Inc. (HB&C) 2011. "*Energy Efficiency Financing in California Needs and Gaps. Preliminary Assessment and Recommendations.*" Presented to The California Public Utilities Commission, Energy Division.

Zimring, M. (2011). "Austin's Home Performance with Energy Star Program: Making a Compelling Offer to a Financial Institution Partner." *Clean Energy program Policy Brief.* Lawrence Berkeley National Laboratory.

End Notes

[1] For example, in California, it is estimated that $70 billion of EE investment in existing buildings will be required over the next decade to achieve the state's policy goals – only a fraction of which will be provided by ratepayer funding (HB&C 2011).

[2] A few examples of this increasing reliance on financing: In California, the Public Utilities Commission has approved $200 million of pilot programs to test whether transitional

ratepayer support can trigger self-supporting programs (CPUC 2013). In Connecticut, the Clean Energy Finance & Investment Authority's 2013-2015 Strategic Plan notes that its programs "will reflect the strategic transition away from technology innovation, workforce development, formal education and subsidies towards a focus on low-cost financing of clean energy deployment...(in order to) seek to leverage ratepayer dollars..."(CEFIA 2013). In New York, the $1 billion Green Bank's goals include overcoming disparate one-time subsidies and offering public credit and investment programs that require only a small amount of government funds (Cuomo 2013).

[3] For example, in California, it is estimated that $70 billion of EE investment in existing buildings will be required over the next decade to achieve the state's policy goals – only a fraction of which will be provided by ratepayer funding (HB&C 2011).

[4] A few examples of this increasing reliance on financing: In California, the Public Utilities Commission has approved $200 million of pilot programs to test whether transitional ratepayer support can trigger self-supporting programs (CPUC 2013). In Connecticut, the Clean Energy Finance & Investment Authority's 2013-2015 Strategic Plan notes that its' programs "will reflect the strategic transition away from technology innovation, workforce development, formal education and subsidies towards a focus on low-cost financing of clean energy deployment...(in order to) seek to leverage ratepayer dollars..."(CEFIA 2013). In New York, the $1 billion Green Bank's goals include overcoming disparate one-time subsidies and offering public credit and investment programs that require only a small amount of government funds (Cuomo 2013).

[5] We describe "offering financing programs" in the broadest sense – this may take the form of direct provision of public or ratepayer capital, direct or indirect support for private sector financial products (e.g., credit enhancement, co-marketing, customer intake), enabling or offering of novel financial products (e.g. on-bill financing) or some combination of these.

[6] Balance sheet treatment refers to whether financing is treated as an "on-balance sheet" or "off-balance sheet" obligation for accounting purposes. "Off-balance sheet" treatment enables non-residential customers to finance EE improvements without increasing their debt-to-equity ratio, a metric which is studied closely by investors and often capped by lenders. For more information on the advantages and disadvantages of off-balance sheet financing, consult an accounting professional.

[7] Other novel financing products may include traditional financing products whose underwriting takes specific account of a property or project's energy efficient features (e.g., *Energy Efficient Mortgages, HUD Powersaver Loans, HUD Green Refinance Plus Multifamily Mortgages*).

[8] Some stakeholders use the term On-Bill Financing to describe on-bill programs that are capitalized with tax payer or utility bill payer capital while using the term On-Bill Repayment to characterize on-bill programs capitalized with private capital. For the purposes of this report, we use the term On-Bill Financing in reference to all on-bill programs, regardless of capital source.

[9] More information on PACE available here: http://www1.eere.energy.gov/wip/solutioncenter/pace.html

[10] More information on OBF available here: http://www1.eere.energy.gov/wip/solutioncenter/onbillrepayment.html.

[11] While novel security may expand customer access to capital, it is important to proceed carefully to ensure that customers are able and willing to repay these novel products as the consequences of non-payment are severe (e.g., foreclosure, utility service disconnection) and high default levels may not be tolerable to policymakers.

[12] Financial product security refers to what "secures" a loan or lease in the event that a customer defaults. For example, home mortgages are secured by the financial institution's right to foreclose on one's home should a customer default on their loan repayment obligation.

[13] These guarantees typically involve complex, long-term contractual arrangements related to ensuring that "baseline" energy use conditions persist in facilities throughout the life of the

guarantee. In commercial buildings, consumers may be reluctant to utilize these contracts due to the loss of long-term flexibility they may face in altering building occupancy, production processes or other factors that may drive their core profitability.

[14] For example, most owners of Class A commercial space and most public entities are deemed creditworthy by private markets and have access to a range of private financing tools (see, for example, Borgeson & Zimring 2013).

[15] It is important to note that this more holistic risk assessment and delivery of capital must be done ca3refully as the consequences of customer financing defaults can be severe and have unintended consequences.

[16] Lower defaults may also be associated with unique features of EE adopters that are not captured in typical creditworthiness analysis, but may increase their predisposition to repay financing.

[17] The U.S. Department of Energy is currently supporting a loan data scoping study to develop best practices EE financing data collection and analysis protocols. Preliminary results will be available in 2014.

[18] Some financial institutions have been motivated to participate in energy efficiency financing pilots primarily for the reputational benefits or for opportunities to cross-sell efficiency customers into other financial products rather than the direct financial returns available from EE financing (Zimring 2011).

[19] Standardization entails consistent financial product origination and servicing protocols, so that a loan or lease originated in California is similar to a loan or lease originated in Oklahoma or New York. This standardization is essential to the process of successfully aggregating and selling these financial products in sufficient volume to attract large pools of low-cost investor capital.

[20] The re-sale of financing products is known as a "secondary" sale as the primary sale is the financial institution's origination of the financial product for the borrower. Financial institutions typically earn fees when they sell financial products to secondary investors.

[21] In the residential sector, many state & local governments used ARRA monies to launch financing programs with local financial institution partners (e.g., credit unions, community banks). Governments typically offered credit enhancements to financial institutions in exchange for interest rate, loan term or underwriting concessions. While many innovative agreements were structured, this innovation came at the cost of standardization. As some programs and their financial institution partners exhaust the money they have available to lend, they have faced challenges in selling their loan pools to "secondary markets" investors due to investor liquidity concerns and lack of historical data to use for loan pool performance modeling.

[22] The Warehouse for Energy Efficiency Loans (WHEEL) model, which relies on a subordinated capital credit enhancement from program administrators, is designed to pool standardized, unsecured residential EE loans for sale to secondary investors.

[23] Several EE financing programs have recently completed or are pursuing a secondary markets transaction (e.g., Pennsylvania's *Keystone HELP* program, New York's *Green Jobs-Green New York* program and Oregon's *Clean Energy Works Oregon* on-bill program).

[24] Residential home performance energy efficiency programs often offer rebates of 25 to 50 percent, yielding between two and four dollars of total EE investment for each rebate dollar expended.

[25] While rebates may deliver limited short-term leverage, utilized as part of market transformation strategies to build customer demand and reduce product costs, these tools may deliver very large long-term leverage.

[26] Many American Recovery and Reinvestment Act (ARRA)-funded EE financing programs targeted residential EE improvements and utilized five to 10 percent loan loss reserves (LLRs). LLRs are a form of credit enhancement that sets aside a limited pool of funds from which lenders or investors can recover a portion of their losses in the event of borrower defaults.

[27] Table 1 provides several examples of program designs that offer program funds/incentives to leverage customer investment in efficiency, ignoring program administration costs.

[28] For example, some individuals and businesses are debt averse or would rather spend available capital on more compelling investments. Some businesses and institutions that have limited or no staff capacity or have already invested in relatively short payback efficiency measures and remaining opportunities have payback times that exceed their preferred internal rates of return on investments. In many cases, the societal benefits (e.g., including environmental externalities) and benefits to utility systems of energy efficiency projects exceed the short-term private benefits to consumers. (Borgeson et al. 2012)

[29] For example, the California Public Utilities Commission (CPUC) has budgeted $8M for utility information technology upgrades to accommodate On-Bill Repayment pilots and $9M of administration and implementation costs as part $75M of EE financing pilots to be operated from 2013-2015 (CPUC 2013).

[30] In a few cases, program administrators are testing whether simply marketing existing market-rate financial products will drive or enable consumer EE adoption at low program cost. There is little data today on the potential efficacy of this lower-cost strategy in driving cost-effective energy savings.

[31] While substantial resources are often targeted to interest rate reductions, the difference in monthly consumer payment on a 10-year, $10,000 loan with 7% vs. 10% interest loan is relatively small (~$16/month), and has uncertain impacts on consumer EE adoption.

[32] These techniques are not, in many cases, mutually exclusive.

[33] The treatment is the intervention that the program is providing to program participants. For example, if a program administrator wants to test whether more customers adopt deep energy improvements if 15 year financing is offered rather than the existing 10 year product, the treatment group would be offered 15 year financing and the control group would be offered 10 year financing.

[34] Where random assignment is not possible or practical, quasi-experimental design techniques assign participants to the two groups in a way that is as close to random assignment as possible. For example, if Town A and Town B share similar demographics, the town lines may form the basis of assigning households or businesses to the treatment or control group.

[35] Random assignment ensures that the customers receiving Program Offer A and Program Offer B are identical *in expectation*. After the randomization has occurred, there will likely be differences that exist between the two groups due to random chance. However, these differences are usually small and statistically insignificant.

[36] For example, if a customer is offered a financing program, and we observe that the customer installs a measure, we cannot then offer the customer a rebate for the same measure in order to compare it to financing.

[37] The exception to this answer is when administrators expect the difference in effectiveness from one offer to another to be extremely high.

[38] For a more detailed description of different types of experimental designs and the analysis needed to evaluate the designs, see: State and Local Energy Efficiency Action Network. 2012. *Evaluation, Measurement, and Verification (EM&V) of Residential Behavior-Based Energy Efficiency Programs: Issues and Recommendations*. http://behavioranalytics.lbl.gov.

[39] For illustrative purposes, we qualitatively assess the relative confidence on can have in these five techniques. While RCT always yields the highest confidence results, depending on the specifics of one's experiment, other techniques may also yield high confidence results. For example if the cutoff point selected is something completely arbitrary, it is likely to result in groups that are as good as randomly assigned. Similarly, if there is a geographic border that runs through a city where there are similar households on both sides of the border, then a geographic location method may yield similar confidence to a cutoff point. The exception is time differences, which is likely to yield lowest confidence results in most cases (See Appendix A for detailed explanation).

[40] Depending on the exact form of the randomization, this may also be called "recruit and delay" or "randomized encouragement design".
[41] This is also called regression discontinuity (RD).
[42] This may be called "difference-in-differences" if data on the outcome being measured is known before the offers are made.
[43] The number of customers needed should be calculated by doing a statistical *power calculation,* and depends on several factors: the research question; the metric used to compare the programs; the experimental design; the minimum difference in program outcomes that would be valuable to learn (e.g., do you need to know if A is 1% better than B, or only if it's 10% better than B?); the percentage of customers that typically decide to do a retrofit after getting any kind of offer from a contractor; the variation in how much money customers spend on retrofits; and other factors.
[44] For illustrative purposes, we qualitatively assess the relative confidence of these five techniques based on what we believe would be a typical implementation approach. While RCT always yields the highest confidence results, depending on the specifics of one's experiment, other techniques may also yield high confidence results. For example if the cutoff point selected is something completely arbitrary, it is likely to result in groups that are as good as randomly assigned. Similarly, if there is a geographic border that runs through a city where there are similar households on both sides of the border, then a geographic location method may yield similar confidence to a cutoff point. The exception is time differences, which is likely to yield lowest confidence results in most cases (see below for a more detailed explanation; basically, because homeowners are likely to invest in energy efficiency measures only once every few years (or longer), an early choice to invest in Option A is likely to preclude a later choice to invest under Option B (even if the homeowner would rather have invested with Option B), and it is therefore harder to compare Offer A to Offer B).
[45] Random assignment ensures that the customers receiving Offer A and Offer B are identical *in expectation*; after the randomization has occurred, there will likely be differences that exist between the two groups due to random chance. However, these differences are usually small and statistically insignificant.
[46] If the contractor knows which customers will receive Program Offer A or B before contacting customers, the contractor may introduce bias by only pursuing leads with customers receiving one offer or the other. The contractor might also present recommendations to the customer in different ways; in which case the results would reflect the preferences of the contractor, not the choices of the customer.
[47] Other evaluation methods require complicated analyses that rely on collecting detailed information on consumers to control for as many consumer differences as possible and isolate the impact of the program element being studied.
[48] Customers closest to the cutoff are the most comparable, while those further away from the cutoff are less comparable. An analysis method called regression discontinuity is the best way to analyze experimental designs based on a cutoff point; it puts more weight on those closest to the cutoff.
[49] To see why this is a problem, consider three cases. In the first case, suppose that one is trying to test which program, A or B, is better. Also suppose that 10 customers would choose to do a retrofit with program A, 15 would choose to do a retrofit with program B, and 25 would decide to do a retrofit with either A or B. If A is offered first and B second, then 35 customers would get retrofits with A, and 15 would get retrofits with program B. One might conclude that A is much better than B, even though in fact program B is better at getting people to do retrofits. If B is first and A second, 40 would get retrofits with B, and 10 with A; one might conclude that B results in 30 more retrofits than A, when in fact it only results in 5 more. In the second case, suppose that A is known to be better than B (e.g., A is 5% financing, and B is 10% financing), and one is trying to test how much better A is. Also suppose that while 50 customers would choose to get retrofits with program A, only 20

would get retrofits with B. If A is offered first and B is offered second, then all 50 customers would get retrofits with program A, and then 0 would get retrofits with program B. One might conclude that A results in 50 more retrofits, when in fact 20 of those people would have chosen retrofits under program B, and so A only results in 30 more retrofits. If A is known to be better than B, and B is offered first and A second, then in our example 20 customers would choose retrofits under program B, and then when the better program A is offered second, 30 additional customers would choose retrofits. Then the additional 30 retrofits could accurately be attributed to program A. In addition, there are other factors that change over time that may affect the way customers react (e.g., changes in the economy, new customers entering the market, changes in interest rates, changes in social culture, etc.).

[50] The number of contractors that would need to be included should be determined using a statistical power calculation; the sample size required may be as much as 100 times that required by randomization of customers. This power calculation will also take intra-class correlation into account and will depend on how many customers each contractor accesses as well as the variance of customers both within and across contractors.

[51] There are many qualified consultants and evaluators that can help set up these designs. This is intended as a primer to familiarize policymakers with the benefits of experimental design and the importance of doing experimental design right.

[52] Census block data is useful if there is a household characteristic that is available in census data that: (a) you believe corresponds to large differences in outcomes, and (b) that the characteristic varies widely across the census blocks that you are studying but does not vary widely within each census block. Household-level survey data is more useful if the characteristic that you would like to test varies widely within census blocks, or you expect small differences in outcomes between the characteristic groups you are testing. For example, it would be useful in the case that you believe that income strongly affects a household's decision to adopt financing over rebates, and that the income of most households within a census block are relatively close to the average income of the census block. Then you could draw conclusions, such as that households in high income neighborhoods tend to choose financing over rebates 20% more often than households in low income neighborhoods.

[53] The number of customers needed depends on several factors and is calculated by doing a statistical power calculation The research question, the metric used to compare the programs, the experimental design, the minimum difference in program outcomes that would be valuable to learn (e.g., do you need to know if A is 1% better than B, or only if it's 10% better than B?), the percentage of customers that typically decide to adopt a retrofit after getting any kind of offer, the variation in how much money customers spend on retrofits and a range of other factors can all impact this calculations.

In: Energy Efficiency Financing Programs
Editor: Louise Altman

ISBN: 978-1-63117-200-7
© 2014 Nova Science Publishers, Inc.

Chapter 2

SCALING ENERGY EFFICIENCY IN THE HEART OF THE RESIDENTIAL MARKET: INCREASING MIDDLE AMERICA'S ACCESS TO CAPITAL FOR ENERGY IMPROVEMENTS[*]

Mark Zimring, Merrian Borgeson, Ian M. Hoffman, Charles A. Goldman, Elizabeth Stuart, Annika Todd and Megan A. Billingsley

> Middle income American households – broadly defined here as the middle third of U.S. households by income – are struggling. Energy improvements have the potential to provide significant benefits to these households – by lowering bills, increasing the integrity of their homes, improving their health and comfort, and reducing their exposure to volatile, and rising, energy prices. Middle income households are also responsible for a third of U.S. residential energy use, suggesting that increasing the energy efficiency of their homes is important to deliver public benefits such as reducing power system costs, easing congestion on the grid, and avoiding emissions of greenhouse gases and other pollutants.

---

[*] This is an edited, reformatted and augmented version of the Lawrence Berkeley National Laboratory, is an excerpt from the report: "Delivering Energy Efficiency to Middle Income Single Family Households.", dated March 6, 2012.

> While middle income Americans have historically invested in improvements that maintain and increase the value of their homes, they have seen an important source of financing – the equity in their properties – evaporate at the same time that their access to other loan products has been restricted. A number of energy efficiency programs are deploying credit enhancements, novel underwriting criteria, and innovative financing tools to reduce risks for both financiers and borrowers in an effort to increase the availability of energy efficiency financing for middle income households. While many of these programs are income-targeted, the challenges, opportunities, and emerging models for providing access to capital may apply more broadly across income groups in the residential sector.

CHALLENGES TO ACCESSING CAPITAL

The upfront cost of comprehensive home energy improvements is a barrier to investment. Many middle income households need financing to overcome this barrier – and capital access has plummeted in the wake of the recession.

Using Home Equity to Finance Home Improvements

Middle income homeowners have historically invested in improving their homes. In 2001, these households accounted for almost a third of all home improvements made in the U.S., and they financed more than 35 percent of their home improvement investments (Guerrero 2003).[1] Compared to other households that financed improvements, middle income households were more inclined than other income groups to finance home improvements by borrowing against housing equity – two thirds of their financing was home-secured (see Figure 1).[2]

This is both good and bad news. The good news is that middle income households have historically invested in home improvements, and many (57 percent) have not needed financing to do so. The bad news is that the recession has eroded household savings – suggesting that more households will need financing to make improvements – at the same time that housing wealth, the primary asset against which middle income households borrow, has declined.

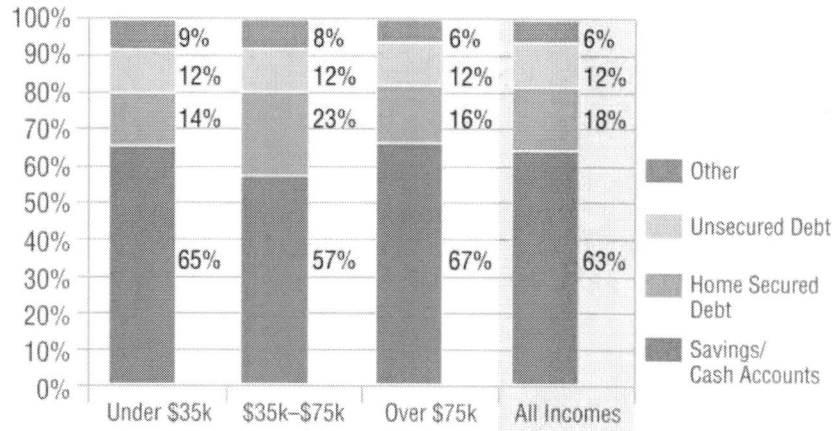

Figure 1. Home improvement financing patterns by income in 2001 (Guerrero 2003).

The Housing Collapse

A number of factors contributed to the enormous speculative housing bubble in the mid-2000s (Lansing 2011). By 2007, primary residences accounted for approximately one third of U.S. household assets. For middle income households, these primary residences represented an even greater share of their assets – almost 50 percent (Bucks 2009).[3] The financial crisis and ensuing recession have since caused a sharp decline in housing values across the United States. Single family home prices have declined by 32 percent from the housing market's 2006 peak and carried household wealth down as well (see Figure 2) (S&P 2011).

This data masks more dramatic regional declines in housing values and the concentration of these price declines in low and middle value properties – those most likely to be owned by middle income Americans.[4] For example, the Case-Shiller Home Price Index indicates that low tier properties in Atlanta have lost 55 percent of their value since peaking at the end of 2006 – almost double the average 23 percent property value decline in the city over that time (see Figure 2).[5,6] In other words, not only did middle income households have more of their wealth invested in their primary residences heading into the recession, but their primary residences have lost a greater percentage of their value than those of their wealthier peers.

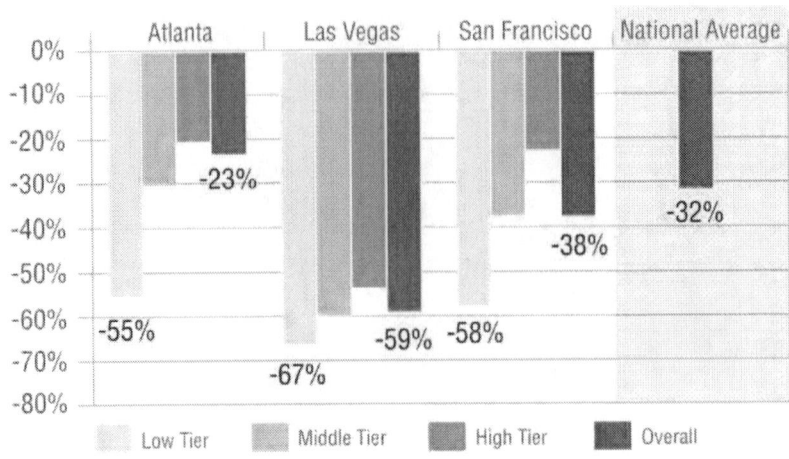

Figure 2. Case-Shiller 20-City Composite Home Price Index of single family home values January 2007 to June 2011 in three major U.S. cities, tiered by initial property value[7] (S&P 2011).

While property values (across tiers) nationally have returned to 2003 levels,[8] it would be incorrect to assume that the housing decline has only set middle income families back eight years. Many homeowners took advantage of rising property values by borrowing aggressively against their growing equity – leaving them with significant debt burdens that are, for some, larger than their home values. In fact, more than a quarter of all single family residential properties (13.3 million households) are now underwater or have near negative equity (<5% equity) (Corelogic 2011). This negative equity is concentrated regionally – the top five states have 38 percent of all negative equity properties.[9] It is reasonable to assume that many of these underwater properties are owned by middle income Americans – these households took on significant debt to purchase and improve properties, are more vulnerable to financial stress during a recession, and lost more of their home's value than their wealthier peers. These underwater households are more likely to behave like renters, under-investing in improving and maintaining their homes.

The news is not all bad though. While a majority of families across income groups have recently experienced declines in income and wealth – and middle income households have been hit harder than their wealthier peers – a large minority of the middle income population has maintained or increased their levels of wealth. From 2007 to 2009, most families (63 percent) experienced wealth declines – for those whose wealth declined, the median loss was substantial, 45 percent (Bricker 2011). However, more than a third of households (37 percent) have not experienced wealth declines or have seen only small changes in wealth. This makes it difficult to make universal conclusions about the state of middle income household finances. While many households are unquestionably suffering – and are likely unwilling or unable to make significant investments in energy efficiency without substantial financial incentives – a large minority of middle income households may be able to invest.

Household Savings & Employment

Many American households feel insecure about their economic futures. Uncertainty about future earnings is high – in 2007, 31.4 percent of all families (across income groups) reported that they did not have a good idea of what their income would be for the next year (Bucks 2009). This uncertainty may well be even higher today as the U.S. unemployment rate has almost doubled since mid-2007. In 2009, almost nine percent of middle income households were unemployed while another 5.5 percent were underemployed (workers that take part-time jobs due to lack of available of full-time jobs) (Sum and Khatiwada 2010).[10]

For those households who have a reasonable expectation of future earnings, the recession has decreased their expectations of annual income growth from around two to three percent before the recession to less than half a percent in its wake – the lowest level in more than 30 years (Dunne and Fee 2011). Lower future earnings expectations are a function of both the recession and longer term trends – over the last 30 years, wages have not kept up with worker productivity gains.[11] Uncertainty and pessimism about future earnings are making households increasingly cautious with their finances as many households report higher levels of desired savings to buffer themselves from economic and other emergencies (Bricker 2011). These homeowners are likely to make fewer proactive home improvements, like energy upgrades, in favor of preserving limited savings and access to credit for unforeseen hardships.

Qualifying for Credit

For those middle income households motivated to pursue energy efficiency, access to low-cost capital is often a significant barrier to investment. Many of the largest energy efficiency loan programs have application decline rates in the 30 to 50 percent range. Household ability to obtain secured financing has declined as housing prices have eroded and lenders have tightened underwriting standards and credit limits (NAR 2011).[12] Similar tightening trends are occurring in unsecured lending as personal creditworthiness has weakened and lenders have responded by increasing the minimum credit scores required to qualify for financing products and reducing the amount of overall credit available to each qualified borrower. Many households turn to high interest credit cards to finance expenditures as their options dwindle. These high-cost financing products are ill-suited to energy improvements – particularly those for which the motivation is to save money – as they worsen the payback period of these investments.

Since 2009, approximately 10,000 households have applied for financing through Pennsylvania's Keystone Home Energy Loan Program (HELP)[13]. About 40 percent of these households earn 80 percent of AMI or less, suggesting that many middle income households are attracted to the program.[14] However, the program's early experience shows that middle income households are more difficult to serve – 57 percent of households earning ≤ 80 percent AMI do not meet the program's underwriting standards compared to 31 percent for households earning >80 percent AMI (see Table 1).[15]

In addition to this higher rejection rate, fewer lower income households move forward with financing than their wealthier peers (58 percent of approved households earning ≤ 80 percent AMI fund loans compared to 73 percent of higher income households) – supporting the idea that, for many reasons, even when financing is available, it is more difficult to motivate middle income households to invest. Still, this data shows some promise as these middle income households account for about a quarter of all Keystone HELP loan volume.

Table 1. Keystone HELP loan application, approval, and loan size rates by income, January 2010 to August 2011. (AFC First)

Household Income	# Applications (% of Total Applications)	Applications Approved (Approval Rate %)	Loans Funded (Approval →Loan Conversion Rate %)	Average Loan Size
<80% AMI	~4,000 (40%)	~1,720 (43%)	~1,000 (58%)	~$7,500
≥80% AMI	~6,000 (60%)	~4,140 (69%)	~3,000 (73%)	~$9,500

Table 2. Changes in VantageScore loan delinquency rates for new loans originated from 2003-2005 compared to loans originated from 2008-2010 (anticipated).[20] (VantageScore)

VantageScore	Loan Delinquency Rate		Delinquency Rate Increase
	2003-2005	2008-2010 (Anticipated)	% increase in rates btw 2003-2005 and 2008-2010
591-610	21.50%	25.44%	*3.9%*
611-630	17.11%	21.18%	*4.1%*
631-650	13.63%	17.81%	*4.2%*
651-670	10.90%	14.62%	*3.7%*
671-690	8.24%	11.74%	*3.5%*
691-710	5.99%	9.74%	*3.8%*
711-730	4.27%	8.11%	*3.8%*
731-750	3.21%	6.64%	*3.4%*
751-770	2.22%	5.28%	*3.1%*
771-790	1.67%	4.29%	*2.6%*
791-810	1.15%	3.33%	*2.2%*
811-830	0.80%	2.57%	*1.8%*
831-850	0.49%	1.78%	*1.3%*
851-870	0.38%	1.40%	*1.0%*
871-890	0.24%	0.90%	*0.7%*
891-910	0.19%	0.63%	*0.4%*
911-930	0.19%	0.53%	*0.3%*

According to the Indianapolis Neighborhood Housing Partnership (INHP), the homeowners that they serve typically have little access to anything but credit card financing – often at annual rates from 15 to 25 percent, so INHP's new EcoHouse Project's mid-single digit fixed-interest rate loans[16] are an attractive tool for enabling energy improvements among households who are otherwise unlikely to be able to access affordable financing. With relatively lenient underwriting standards including credit scores as low as 580,[17] INHP is able to accommodate a wider range of applicants.[18]

Credit scores estimate an individual's likelihood of repaying certain types of debt relative to one's peers. Credit scores are a key metric for most lenders in evaluating consumer creditworthiness. Because credit scores are relative measures, a large shift in bill payment trends, like that caused by the recession, has triggered an increased likelihood of loan default for each "band" or range of credit scores. In other words, a credit score of 720 today reflects a higher estimated risk of loan non-payment than a credit score of 720 in 2005. For example, in the case of VantageScore,[19] the delinquency rate on a new loan issued to a person with a 720 score between 2008 and 2010 is expected to be twice as high as on a new loan issued between 2003 and 2005 (see Table 2).

Although credit scores do not explicitly take income into account, middle income households are likely to have lower credit scores than their wealthier peers (see Figure 3). These lower scores may be in part due to creditworthiness and in part due to the way in which scores are calculated, notwithstanding issues about how middle income households manage their credit. For example, a key factor in calculating credit scores is one's ratio of credit utilization to credit availability – many middle income households have less overall credit availability than their wealthier peers, often causing their credit utilization rate to be higher and their credit scores to be lower. This lower credit access may be a function of many things, including lower absolute levels of home equity and post-recession reductions in the maximum loan sizes lenders offer to customers. In other words, income implicitly impacts some credit scores – even in cases of identical loan repayment histories, middle income households may be assigned lower credit scores than their wealthier peers.

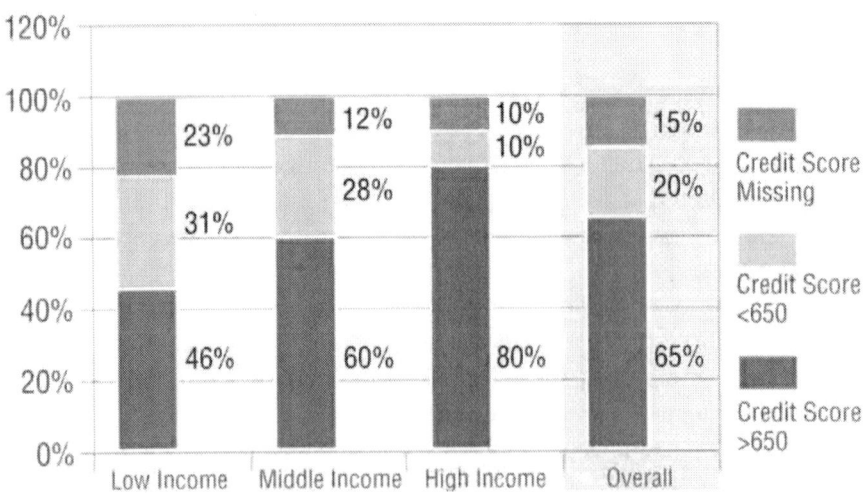

Figure 3. Homeowner credit scores above and below 650 by income in Q4 2010[21] (Energy Programs Consortium).

Scaling Energy Efficiency in the Heart of the Residential Market 61

Most lenders use credit scores as just one of several metrics for evaluating consumer creditworthiness. Underwriting standards for loan products, including those for home improvements, frequently include both a minimum credit score and a maximum debt-to-income (DTI) ratio.[22] A Federal Reserve Board study found that more than 20 percent of all households with home-secured debt had net DTI ratios higher than 40 percent, suggesting that as many as one in five households may not qualify for financing programs that include a maximum DTI underwriting requirement (Bucks 2009).[23] These numbers are higher among middle income households – more than one in three middle income households (35 percent) had net DTIs exceeding 40 percent.[24]

Program experiences to date suggest that maximum DTI underwriting requirements are significant barriers to capital access. For example, NYSERDA has declined more loan applications because household DTI ratios exceed the allowable limit than for any other reason. Forty-three percent of NSYERDA's loan application declines (17 percent of loan applicants) have been caused by excessive DTI ratios while just 23 percent of declines were triggered by low household credit scores (See Figure 4). Major credit events like bankruptcy, foreclosure, repossession and outstanding collections account for more loan denials (33 percent) than low credit scores – these loan applicants will be very difficult to serve moving forward.

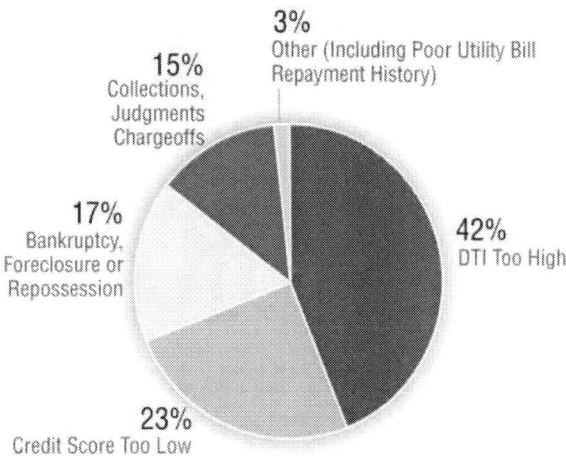

Figure 4. Reasons for application rejection in NYSERDA's residential energy efficiency loan program November 2010-October 30, 2011 (NYSERDA).

Table 3. Credit score and corresponding delinquency projections. (Transunion 2011 in SEE Action Financing WG)

FICO Score Range[25]	Delinquency Projection (% Likelihood)
300-499	87
500-549	71
550-599	51
600-649	31
650-699	15
700-749	5
750-799	2
800-850	1

OPPORTUNITIES FOR INCREASING ACCESS TO CAPITAL

Middle income households clearly need new ways of accessing affordable credit if they are to make home energy upgrades. However, it is important to acknowledge that there can be negative consequences to promoting loans and other products to particularly vulnerable segments of the population. Especially if programs are not ensuring savings, care needs to taken with regard to who is given access to credit and what claims are being made about the benefits of energy improvements.

Underwriting criteria exist for a reason – to ensure that those that get access to financing are willing and able to make required monthly payments. For credit scores, the majority of middle income homeowners (60 percent) have scores of 650 or higher. For those with scores below 650, default risk skyrockets – the projected delinquency rate on unsecured loans more than doubles from 15 to 31 percent for individuals with FICO scores from 600-650 compared to their peers in the 650-700 score band (see Table 3).[26] This raises important questions about how to expand energy efficiency financing – particularly in the absence of certainty that the dollar value of energy savings will be sufficient to cover the full cost of the improvements over the measure's expected lifetime. Debt to income constraints raise similar issues – households with high DTIs are unlikely to have significant cash flow buffers at their disposal should energy improvements not deliver sufficient energy bill reductions to offset financing costs.

With those precautions acknowledged, there are ways that capital can be made more accessible and affordable in appropriate ways, and with prudent safeguards. This section describes options for using credit enhancements, alternative underwriting criteria, and other financing mechanisms that might better serve middle income households.

Credit Enhancements

By reducing lender risk, publicly-supported credit enhancements can leverage these limited public monies and attract additional capital for residential loans.[27] Credit enhancements are used to reduce a lender's risk by sharing in the cost of losses in the event that loans default. These enhancements can take the form of loan loss reserves (LLRs), subordinated debt, and guarantees.[28] LLRs, often funded with ARRA or utility-customer funds, are the most commonly used credit enhancement, and they are frequently deployed to reduce borrowing costs or extend borrowing terms for program participants that would likely qualify for other (more expensive) loan products. Rather than simply lowering interest rates, a few innovative programs are using credit enhancements to incentivize their financial partners to offer energy improvement loans to households who would otherwise not have to access capital. Indianapolis is using a large LLR – with 50 percent[29] of losses covered – to households in its target income demographic,[30] and the cities of Madison and Milwaukee used part of their DOE Better Buildings grant to structure a $3 million LLR to expand access to their loan product. This five percent loss reserve reduces the lender's losses in the event of loan defaults and supports a loan pool of up to $60 million. It has been structured so that the cities' financial partner, Summit Credit Union, can recover more funds from the LLR on each loan default for lower credit quality consumers. Typically, a lender must absorb a fixed portion of each loss from any single loan to ensure it is appropriately motivated to lend responsibly. By allowing lenders to collect a greater percentage of their loss on loans to customers with low credit scores, the two cities were able to lower the minimum qualifying credit score to 540 – well below typical loan product eligibility (see Table 4).

Table 4. Milwaukee/Madison-Summit Credit Union loan loss reserve agreement. (Wisconsin Energy Conservation Corporation)

FICO Score Range	% of Each Loss Covered By LLR	% of Each Loss Absorbed by Credit Union
690+	70%	30%
650-689	80%	20%
610-649	90%	10%
540-610	95%	5%

One issue that this type of arrangement raises is whether the lender will continue to be appropriately motivated to responsibly underwrite loans. In the Milwaukee/Madison case, this concern is mitigated by Summit Credit Union's demonstrated commitment to responsible lending to low and moderate income households. Summit's Chief Lending Officer, Dan Milbrandt, pointed out that expanding access to financing is difficult and that it takes effort on the part of the credit union to understand applicants' credit situations and figure out where, on the margin, less creditworthy households are willing and able to take on debt. "You have got to be willing to move beyond automated underwriting. There is a gray area, and Summit has experience examining mitigating factors so that we can responsibly lend to less credit qualified customers."

Alternative Underwriting Criteria

Rather than using credit enhancements to expand financing to "riskier" borrowers, a number of energy efficiency financing programs are deploying alternative underwriting criteria to identify creditworthy borrowers that do not meet traditional lending standards. NYSERDA's recently-launched Green Jobs-Green New York (GJGNY) initiative is using a Two-tiered underwriting process to expand access to financing for its Home Performance with ENERGY STAR© (HPwES) program.[31] Tier One underwriting uses standard credit score (minimum 640)[32] and DTI (maximum 50 percent) metrics to evaluate creditworthiness; 48 percent of applicants are rejected for this financing. NYSERDA is trying to reduce this decline rate with its Tier Two standards that offer households with low FICO scores or high DTIs a second opportunity to qualify for GJGNY financing (see Table 5 for a description of

Tier Two underwriting standards). For those households with FICO scores below 640, NYSERDA Tier Two standards increase the maximum DTI to 55 percent and use utility bill repayment history in lieu of credit score to assess creditworthiness. For households with a FICO score above 680 that were rejected from Tier One because they had a DTI ratio above 50 percent, Tier Two standards increase the maximum DTI to 70 percent and use utility bill repayment history.[33]

Since its November 2010 launch, over $7.8 million has been loaned to 908 households through the GJGNY initiative, of which 48 loans ($417,888) have been issued to households qualifying under the new Tier Two standards. Tier Two underwriting criteria have increased access to capital on the margin, increasing NYSERDA's overall loan application approval rate by over two percent. This increase may underestimate the impacts of using utility bill repayment history as a means of assessing creditworthiness – a multi-step application process appears to have been a significant hurdle for many potential Tier Two participants and NYSERDA only launched the "High DTI" underwriting criteria in July 2011[34] (See Figure 5 for a summary of NYSERDA's GJGNY loan application data).

NYSERDA has already made several changes to the Tier Two underwriting criteria since the initiative launched in 2010, which is indicative of the flexibility that is essential to experiment with increasing access to financing. One key challenge has been gaining access to customer utility bills for Tier Two consideration. Many programs around the country have struggled to access customer utility bills. In NYSERDA's case, better access to utility billing information is important to deploying alternative underwriting criteria.

Table 5. New York's Green Jobs-Green New York financing underwriting criteria. (NYSERDA)

Eligibility Requirements		Participant Benefits
Tier 1 FICO≥640 DTI≤50%		3.99% financing Up to $25,000 (3.49% with Automated Clearinghouse (ACH) payment)
Tier 2 (Problem = Low FICO) FICO≤640 DTI≤55% Strong Utility Bill & Mortgage Repayment History	Tier 2 (Problem = High DTI) FICO≥680 50≤DTI≤70% Strong Utility Bill & Mortgage Repayment History	

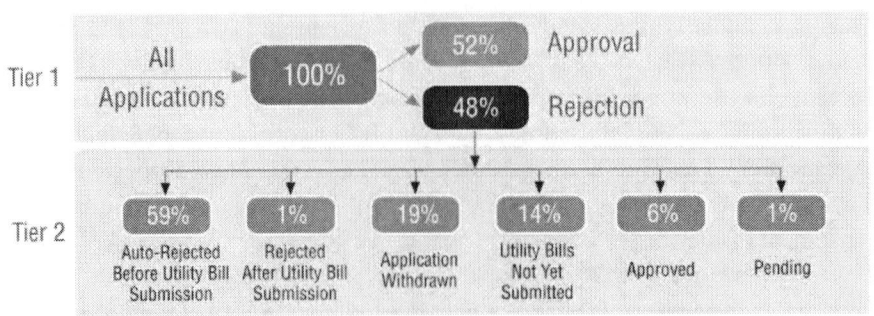

Figure 5. Summary of NYSERDA's GJGNY loan application process and data (November 2010 to December 2011) (NYSERDA).

Other programs, including Midwest Energy and Clean Energy Works Oregon (CEWO), also use utility bill repayment history to evaluate creditworthiness. CEWO's underwriting process is notable for its low cost – while it includes a credit score check, instead of analyzing an applicant's DTI, CEWO examines utility bill repayment history. Using utility bill repayment history in lieu of DTI's significantly reduces loan underwriting expenses, and because more households in many programs are rejected for financing due to high DTIs than low credit scores, it may be an effective approach. The early data are promising – CEWO's application decline rate is just 10 percent since the program's 2009 launch – well below that of other energy efficiency loan programs. CEWO's financing partner, Craft3 (formerly known as Enterprise Cascadia), has dispersed dispersed $14.7 million for 1,180 loans as of January 31, 2012.[35]

These initiatives are relatively new, so it is too early to draw firm conclusions about whether these criteria will be effective at identifying households who can afford to take on debt to invest in energy improvements.[36] While there is reason for some skepticism about the predictive power of utility bill repayment history on loan performance,[37] if on-time utility bill payment turns out to be a good borrower risk assessment tool, it has the potential to increase financing access – and is especially appealing if loan repayments are made on the utility bill as the CEWO program offers. Using on-bill repayment is likely to reduce loan delinquencies, especially where nonpayment can result in disconnection (which is not the case for CEWO).

Innovative Financing Tools

In addition to making standard loan products more accessible, a number of new financial products may be more effective at serving middle income households. Here, we highlight four of these financing tools: OBF loan products that are paid off when properties transfer, employer-offered financing that is deducted from paychecks, and property assessed clean energy (PACE).

On-Bill Financing (OBF)

On-bill financing is a tool through which a customer's utility bill is used to collect loan payments for energy improvements. Utilities or third parties can provide the up-front capital for the energy upgrades and the loan can be structured as an unsecured consumer loan, a secured loan, or can be attached to the meter (as opposed to the individual).[38] Some utilities have expressed reservations about performing lending functions in-house, suggesting that third party-funded on-bill models in which financial institutions have core lending responsibilities (e.g. managing credit risk, hedging interest rate risk) and utilities manage customer interactions (e.g. demand creation, quality assurance).

Because many households have long histories of paying their utility bills regularly, some financial experts believe that on bill repayment will reduce loan delinquency. On-bill financing for energy improvements is the most integrated with the savings those improvements are expected to deliver – which may help to alleviate consumer reluctance to take on debt to pay for them. Midwest Energy in Kansas operates a meter-attached residential loan program. If an individual doesn't pay their bill and leaves the property, only the late payments at that point are uncollectible. Any remaining monthly payments transfer to the next customer at that meter. Over three years, the Midwest Energy program has issued about 600 loans for a total of more than $3.3 million in funding, and to date less than one percent of loans have been uncollectible (in line with the uncollectible rate of their other utility revenue).

Loan Products that Are Paid off when Properties Transfer (Deferred Loans)

Some middle income households simply do not have the financial capacity to make consistent principal and interest payments on debt. This is especially true when the financed improvements lead to uncertain cash flow, or if building rehab needs to be funded in addition to energy upgrades, increasing net monthly payments. There are many housing and economic development agencies around the country that will fund home improvements through deferred loans – often health and safety-related rehab for fixed income seniors that have equity in their homes. No monthly payments are required, but a lien is attached to the property that must be paid off when the property is sold or otherwise transferred.

The Opportunity Council in Washington uses these deferred loans for repairs needed before free weatherization services to low income families. In Camden, New Jersey the city is using Recovery Act funds to create a revolving loan fund to offer residents a home energy upgrade, paid for with a deferred loan. The Wyoming Energy Savers (WES) loan program offers both amortized and deferred loans based on participant income.[39] Those households earning less than 50 percent of AMI qualify for deferred loans, while those households earning 50-80 percent of AMI qualify for amortizing loans.[40] Income-qualified households who are current on their mortgage are eligible for loans up to $15,000 for a list of pre-approved measures including heating equipment and weatherization measures. Deferred loans are offered at 3 percent interest due at time of home property transfer or sale.[41] One key disadvantage to this product type is that borrowed funds are likely to revolve very slowly.

Paycheck-Deducted Loans

Paycheck-deducted financing involves repaying a loan through regular, automatic deductions from an employee's post-tax paycheck. The Clinton Climate Initiative (CCI) is piloting a program called the Home Energy Affordability Loan (HEAL) in Arkansas,[42] which allows employees of participating companies to finance energy upgrades with repayment through a payroll deduction. Originally, the model entailed CCI providing technical assistance for companies to make energy efficiency improvements to their own facilities. These companies would then put a portion of the savings from these improvements into a revolving loan fund for employees. The employer-assisted model is still available, but CCI found that employee demand for financing was larger than the energy savings companies were realizing, and some companies have policies that preclude lending to employees. CCI developed a second model in partnership with local credit unions, in which a credit union, rather than the employer, provides the loan capital and loan repayment is deducted through payroll and automatically transferred to the credit union. For one pilot with the largest hospital in Arkansas, the hospital's credit union is offering 5.75 percent interest for up to three years for unsecured loans to employees who have worked at the hospital for at least three years. The loans are unsecured, but the payroll deduction allows the credit union to do lighter underwriting and offer a lower interest rate than they would otherwise offer for standard unsecured loans.[43] Beyond this security, some experts believe that households may be more likely to pay these loans because they are offered through — or are supported by — their employer, and they want to be seen as responsible employees and members of the company's social community.

Property Assessed Clean Energy (PACE)

For those middle income households who have equity in their homes, PACE may be a promising financing tool if it gets past the current regulatory hurdles. PACE programs place tax assessments in the amount of the improvement on participating properties, and property owners pay back this assessment on their property tax bills. Like other property taxes, these assessments are treated as senior liens – which makes them very secure. PACE is debt of the property, which suggests that underwriting need not be based on a borrower's personal creditworthiness (and that the financing can be transferred with the property) – potentially getting around the credit score and debt-to-income issues highlighted in this chapter. Residential PACE currently faces significant regulatory hurdles, which have largely eliminated its use around the country, pending court rulings or federal legislation.[44]

Loan Pool Aggregation versus Loan Pool Separation

As energy efficiency markets scale, and billions of dollars of private capital become necessary to meet household demand, program administrators and/or their financial partners will likely need to sell energy efficiency loans to "secondary market" purchasers.[45] One important issue to consider as energy efficiency financing markets scale is whether, before being sold into secondary markets, pools of loans made to lower credit quality households should be separated from pools of loans issued using "conforming" underwriting standards to higher credit quality households.[46] Some experts suggest that blended pools of loans, in which strong credits mitigate the risk of weaker credits, will be necessary to deliver attractive loan capital to middle income households at scale. These experts argue that credit enhancements should be deployed to reduce investor risk until a sufficient data set has been accumulated to evaluate the risk of these blended pools.

Others suggest that separate pools are more appropriate, because conforming loan pools would be easier to sell into secondary markets and because these pools would attract the lowest-cost capital available – enabling programs and financial institutions to pass on low-cost financing to these higher-credit households. They suggest that less creditworthy households should be offered public funding or that their loans should be heavily credit-enhanced if sold to private investors. The path forward may, ultimately, be a function of what risks secondary market investors are willing to bear, and whether policymakers deem the credit enhancements necessary to incentivize greater risk-taking to be a reasonable use of limited public monies. Today, it is not clear that demand is at the requisite scale that developing secondary market access should be a national priority. Local, often socially-interested financial institutions (e.g. credit unions, CDFIs, coops) are often offering more attractive loan terms to customers than regional and national lenders (and holding these loans on their balance sheets).[47]

REFERENCES

Bricker, J., Bucks, B., Kennickell, A., Mach, T. & Moore, K. (2011). *Surveying the Aftermath of the Storm: Changes in Family Finances from 2007 to 2009*. Federal Reserve Board Finance and Economics Discussion Series. Federal Reserve Board, March 2011. LINK

Bucks, B. K., Kennickell A. B., Mach T. L. & Moore K. B. (2009). *Changes in U.S. Family Financings From 2004- 2007: Evidence from the Survey of Consumer Finances. Federal Reserve Board Federal Reserve Bulletin*, vol. 95, A1-A55. LINK

Corelogic. (2011). *New Corelogic Data Reveals Q2 Negative Equity Declines in Hardest Hit Markets and 8 Million Negative Equity Borrowers Have Above Market Rates*. Press Release. LINK

Dunne, T. & Fee, K. (2011). *Economic Trends* Federal Reserve Bank of Cleveland. June 2011. LINK

Guerrero, A. M. (2003). *Home Improvement Finance: Evidence from the 2001 Consumer Practices Survey*. Joint Center on Housing Studies, Harvard University. LINK

Lansing, K. J. (2011). *Gauging the Impact of the Recession*. Economic Letter, Federal Reserve Board of San Francisco. LINK

National Association of Realtors (NAR). (2011). *Existing Homes Sales Rise in March*. Press Release. LINK

Sum, A. & Khatiwada, I. (2010). *Labor underutilization problems of U.S. workers across household income groups at the end of the great recession: a truly great depression among the nation's low income workers amidst full employment among the most affluent*. Center for Labor Market Studies Publications. Paper 26. LINK

Standard & Poors (S&P). (2011). *Seasonally-adjusted S&P/Case-Shiller Home Price U.S. National Index Level Q2- 2011*. LINK

Transunion. (2011). *Sample FICO Score Distribution*. In SEE Action Financing Solutions Working Group. LINK

U.S. Department of Energy Technical Assistance Program Webcast (TAP Webcast). (2011). *CDFIs: Opportunities for Partnerships with Energy Efficiency Programs*. LINK

End Notes

[1] In 2001, middle income households spent an average of $8,700 when using home-secured financing to pay for home improvements (Guerrero 2003). The level of home improvement spending impacted homeowner financing patterns. For improvements of $5,000 to $20,000, middle income households used home secured financing for 22% of expenditures, less than their overall average, but 10% more than their wealthier peers for the same expenditure range (Guerrero 2003).

[2] Home-secured financing includes home equity loans, home equity lines of credit and cash out refinancing. Unsecured financing includes unsecured loans and credit cards.

[3] The Federal Reserve Board data uses percentile of income. We use the 40th-70th percentiles ($29,680 to $79,100) to approximate middle income. In 2007, the overall average primary residence asset value as a percentage of wealth was 31.8 percent across all income groups, versus 48.4 percent for middle income households.

[4] The median middle income home value in 2007 was $150,000 (U.S. Census). Assuming a value decline of approximately one third, this median value is likely to be approximately $100,000 today. This value falls into the low tier of the 3-tiered Case-Shiller housing value pricing index across all of the index's 20 major metropolitan statistical areas (MSAs) except for Phoenix (where properties under $95,901 are in the low tier).

[5] In Atlanta, as of June 2011, low tier properties are those valued under $130,356, middle tier are those valued $130,357-$241,832 and high tier are those valued over $241,832.

[6] Case-Shiller Seasonally-Adjusted Home Price Tiered Index Data. June 2011.

[7] *Ibid*. In Las Vegas, Low Tier properties are those valued under 118,226, Middle Tier are $118,226- $178,664 and High Tier are those valued over $178,664). In San Francisco, Low Tier properties are those valued under $325,457, Middle Tier are $325,457-$601,276 and High Tier are those valued over $601,276.

[8] *Ibid*.

[9] *Ibid*. The top five states are Nevada (60 percent underwater), Arizona (49 percent underwater), Florida (45 percent underwater), Michigan (36 percent underwater) and California (30 percent underwater).

[10] As of Q2 2011, the unemployment and underemployment rates have dropped by approximately 0.5 percent across income groups.
[11] For a detailed discussion on wage stagnation, visit the Employment Policy Research Network: http://www.employmentpolicy.org/sites/www.employmentpolicy.org/files/field-content-file/pdf/Mike%20Lillich/EPRN%20WagesMay%2020%20-%20FL%20Edits_0.pdf
[12] Requirements to obtain conventional mortgages have been tightened, with the average credit score rising to about 760 in the current market from nearly 720 in 2007; for FHA loans the average credit score is around 700, up from just over 630 in 2007.
[13] Keystone HELP offers unsecured loans and loans secured by a subordinate lien mortgage at various interest rates. The specific offering depends on the measures financed and loan size. Underwriting includes a minimum credit score of 640, no bankruptcy, foreclosure or repossession in the last seven years, no outstanding collections, judgments or tax liens exceeding $2,500 and a 50 percent maximum DTI.
[14] 80 percent State Median Income (SMI) in PA is $39,600 – suggesting that despite variance of AMI across regions in the U.S., many households who apply for Keystone HELP meet our middle income definition.
[15] Program underwriting is based on these criteria: Minimum FICO Score 640; no Bankruptcy, Foreclosure, Repossession in past seven years; no Unpaid Collection Accounts, Judgments, Tax Liens >$2,500.
[16] Loan interest rates are based on U.S. Treasuries. In July 2011, interest rates on secured loans were 5.97 percent and on unsecured loans were 6.66 percent.
[17] Households with credit scores as low as 580 can qualify for secured financing through INHP's EcoHouse Project loan program. Most national lending products require a minimum credit score of 640 to 680.
[18] For more information on the Indianapolis Neighborhood Housing Partnership EcoHouse Loan Program, see the Policy Brief posted here: http://middleincome.lbl.gov/
[19] VantageScore is a one of a number of consumer credit risk scores that use credit data and analytics as one measure of consumer creditworthiness. Many score models exist in the marketplace (others, like Fair Isaac (FICO) are mentioned elsewhere in this report). However the score values from one model are not comparable to the values of other score models – that is, a 650 score from one model is not comparable to a score value of 650 from a different model.
[20] Credit score models, including the VantageScore model, do not predict absolute delinquency rates. Rather, these models predict the "likelihood" of default for each consumer whose score falls within the indicated range.
[21] Due to data limitations, for the purposes of the credit score analysis we use household income of $30,000 to $70,000 to define middle income. Credit score data from Energy Programs Consortium; based on analysis of TransUnion credit data from Intellidyn.
[22] The debt-to-income (DTI) ratio is a measure that reflects a household's ability to service its existing debt with current gross income. A household with a DTI ratio of 50 percent has annual debt service payments that equal 50 percent of the household's annual gross income. A maximum DTI is intended to ensure that borrowers have sufficient cash flow to make loan interest and principal payments.
[23] The Federal Reserve Board study's net DTI ratio calculation is not directly comparable to the way in which energy loan programs calculate DTIs. This calculation considered income net of taxes while loan underwriters use gross (e.g. before tax) income. These numbers may, therefore, overstate the problem. However, middle income households typically face lower effective tax rates than their higher income peers, suggesting that the gap between middle and higher income households with excessive DTI ratios may be larger than these numbers show.
[24] This includes both owners and renters.
[25] These scores are not directly comparable to the VantageScore scores previously referenced, due to different credit calculation methodologies.

[26] One reason for this significantly higher default rate among lower credit score customers may not be lack of creditworthiness, but instead that these households are only offered high interest rate loan products that are more difficult to pay off.

[27] Loan loss reserves (LLRs) (see next footnote) reduce lender risk by providing first loss protection in the event of loan defaults. For example, a 5 percent LLR allows a private lender to recover up to 5 percent of its portfolio of loans from the LLR. A $20 million fund of private capital would need a $1 million public LLR (5 percent coverage), leveraging each public dollar 20 to 1. On any single loan default, the LLR often pays only a percent of the loss (often 80 percent) to ensure the lender is incentivized to originate loans responsibly.

[28] Loan loss reserves are held in an account and protect a lender against a specific level of loan losses. Subordinated debt stakes are similar to LLRs – instead of being held in an account, subordinated debt is lent out to customers, and the subordinated debt stake absorbs all losses up to a specified level. Loan guarantee protection can vary depending on the agreement, but can cover all or part of a lender's losses.

[29] In comparison, most LLRs for Recovery Act-funded programs have covered 5 to 10 percent of a portfolio's losses.

[30] INHP is targeting 80 percent of its EcoHouse lending to households at or below 80 percent of AMI and the remaining 20 percent to households earning between 80 percent and 120 of of AMI. 120 percent of AMI for Indianapolis household of four is $79,200.households and 80% AMI for an Indianapolis household of four is $52,800.

[31] Households earning less than 80 percent of AMI are eligible for NY's AHPwES program, which provides a 50 percent rebate up to $5,000.

[32] Minimum FICO score is 640, unless self-employed – minimum 680 if self-employed for at least 2 years, or minimum 720 if self-employed less than two years.

[33] There are many ways to calculate debt to income (DTI) ratios. Most programs use gross income. It is not clear, therefore, that a 70 percent DTI maximum is a meaningful metric for assessing creditworthiness (e.g. many households pay close to a third of gross income in taxes, suggesting that this metric might exclude very few households as debt service could include 100 percent of household net income). NYSERDA already assesses DTI ratios as part of its Tier 1 evaluation, but programs considering a different underwriting process should consider this issue.

[34] GJGNY requires that applicants not qualified under Tier One but not initially disqualified from Tier Two for reasons unrelated to utility bill repayment history (e.g. recent bankruptcy, high DTI) to proactively submit utility bills. This step has been a barrier as more than 80 percent of applicants have failed to follow-up with bill submission. While the overall loan application approval rate increased by just 2.6 percent, this may underestimate the impacts of using utility bill repayment history as other underwriting criteria and the multi-step application process appear to be barriers. For example, if 84 percent (the rate of loan approval for applicants that submitted utility bills) of all households not automatically disqualified from the Tier Two track (e.g. those that failed to submit their utility bills) had been approved, GJGNY's approval rate would have increased by 16 percent.

[35] Thus far three loans have defaulted totaling $39,674 in charge-offs. Their current criticized assets equal 3.87 percent of the outstanding portfolio, including watch list assets at 2.89 percent and problem assets at 0.98 percent. However, it is also important to note that most applicants – both those declined and those approved – have strong credit scores, most above 700.

[36] Ultimately, the viability of these alternative underwriting approaches must be assessed not based on how many loans additional loans are made, but whether such loans exhibit payment performance that justifies approving borrowers who would otherwise not qualify for financing.

[37] Households are uniquely motivated to pay utility bills to ensure that their power stays on. This motivation may not hold for unsecured loans, where the penalty for non-payment is a credit score reduction.

[38] If the repayment obligation is attached to a household's utility meter (meter attached), the obligation to pay the loan can stay with the property if a tenant or homeowner moves. In some programs, nonpayment of the bill can trigger utility shut-off of service, a powerful customer incentive to make interest and principal payments. [38] Because of this enhanced security, a household's credit characteristics become less importing to underwriting. However, the same consumer protections that guard against utility service cancellation in the event of utility bill nonpayment also protect on-bill financing borrowers from meter shutoff in the event of loan nonpayment. Some utility commissions have expressed support for facilitating the convenience and messaging of on-bill repayment but are not inclined to support meter attachment which could lead to service disconnection. The extent to which meter-attached financing might influence real estate transactions properties also remains an open question.

[39] An amortizing loan is one in which loan principal is paid down over the course of the loan. A deferred loan is one in which principal and/or interest payments are postponed for a specific period of time or until a specific trigger (e.g. property transfer).

[40] Depending on the county, 50 percent of AMI ranges from $33,700 to $47,450 for families of 4, and 80 percent of AMI ranges from $53,900 to $64,200.

[41] For more information, visit http://www.wyomingcda.com/files/WESDes.pdf

[42] The Clinton Climate Initiative plans to replicate the program in other states beginning in 2012. More information on the program is available here: www.clintonfoundation.org/what-we-do/clinton-climate-initiative/cci-arkansas.

[43] In some states, a direct lender or employer deduction from the paycheck may not be legal as employees must maintain personal control over their income. These states include: Illinois, Indiana, New Hampshire, New Jersey, New York, Washington, D.C. and West Virgina. However, this is generally viewed as a technical obstacle, and customers may voluntarily setup automated paycheck allocations to personal accounts, which are then automatically transferred to lenders or employers

[44] For more information, visit http://www1.eere.energy.gov/wip/pace.html

[45] A secondary market is a market into which previously issued financial instruments (e.g loans, stocks, bonds) can be sold.

[46] A conforming loan is a loan whose structure (e.g. security, term) and underwriting criteria (e.g. minimum credit score) meet specific guidelines. The bellwether of conformity for energy efficiency loans is the Fannie Mae Energy Loan.

[47] These financial institutions often see energy efficiency lending as serving their social missions. In addition, efficiency lending often offers them a low-cost marketing tool, which warrants attractive lending terms. In Austin, Texas, Velocity Credit Union approved, funded and cross-sold energy efficiency loans at a higher rate than its other lending products. For more information, visit LBNL's policy brief on Austin Energy's Home Performance with ENERGY STAR© program: http://eetd.lbl.gov/ea/emp/reports/ee-policybrief_032211.pdf

In: Energy Efficiency Financing Programs
Editor: Louise Altman

ISBN: 978-1-63117-200-7
© 2014 Nova Science Publishers, Inc.

Chapter 3

THE LIMITS OF FINANCING FOR ENERGY EFFICIENCY[*]

Merrian Borgeson, Mark Zimring and Charles Goldman

ABSTRACT

Financing is an appealing concept when efficiency program budgets are a small fraction of the overall level of efficiency investment needed to achieve our public policy goals – but that does not mean financing is always the solution, and it is certainly not the *only* solution. We show that financing can, in some cases, increase the leverage of public dollars. In most cases, however, it is not able to drive demand to the same degree as direct incentives like rebates and so cannot be expected to replace other incentives in the current marketplace. We also show that subsidized financing for those who already have access to capital may be a poor use of public funds, and that increasing access for those who are currently underserved will likely require ongoing subsidy. This is not to say that financing is unimportant – financing is one of many important tools for scaling efficiency and should be employed thoughtfully with the questions outlined in this paper in mind.

Keywords: Energy Efficiency Financing, Energy Efficiency Programs, Financial Incentives for Energy Efficiency, Retrofits, Efficiency Programs, Incentives

---

[*] This is an edited, reformatted and augmented version of the Lawrence Berkeley National Laboratory, dated August 2012.

INTRODUCTION

States and utility regulators are increasingly adopting aggressive energy efficiency targets for existing buildings. To achieve those goals, utilities and governments are increasing their reliance on programs that improve the energy efficiency of the entire building, instead of focusing on single measures or end uses (e.g., lighting). These more comprehensive programs typically require customers to pay a significant portion of the improvement costs. In this environment, financing has been put forward as a tool that can drive investment in comprehensive improvements where the energy savings yield cash flows in excess of loan interest and principal payments.

This "financing is the solution" view is reinforced by the negative cost bars for many efficiency improvements on the McKinsey cost of carbon abatement curve, and the refrain that efficiency is the "low hanging fruit" or even the "fruit on the ground". The narrative is attractive to program administrators and state regulators concerned about the potential short-term impact on utility rates of meeting aggressive energy efficiency targets. It is also attractive to policymakers struggling with the reality that program budgets are a small fraction of the overall efficiency investment needed to achieve our public policy goals (e.g. reducing the cost of serving energy consumers, easing congestion on the grid, minimzing environmental impacts, equitable access to efficiency opportunities). While this idea – that financing can deliver the long-heralded low hanging fruit of energy efficiency in buildings – is intellectually appealing, financing as the most important element of program design strategy has not been widely substantiated in over 25 years of experience with financing programs.

The reality is far more complex. There are individuals who are debt averse, or can't (and often shouldn't) qualify for credit, or would rather spend available capital on more compelling investments. There are businesses, governments, and institutions that have no debt capacity, or that have replaced their lighting already and aren't interested in efficiency investments with more than a two or three year payback, or that don't have staff available to manage the work. In some regions of the country the lowest hanging fruit has already been plucked. In other regions the climate or low energy prices make the case for aggressive efficiency more challenging without a long-term view that considers efficiency's overall benefits, public and private. This challenge is magnified in some regions where there isn't a trained workforce and a developed energy efficiency services sector to provide an attractive package of measures. Even for those motivated to invest in efficiency, the transaction costs of making these improvements can be high.

While energy efficiency is often the lowest-cost energy resource, and financing is an important tool for enabling efficiency, the focus on financing by policy makers, program administrators, and advocates is often out of scale with what financing can be expected to accomplish – and certainly out of scale with what financing has accomplished to date (Bell, Nadel & Hayes 2011; Brown & Conover 2009; Fuller 2009; Palmer, Walls & Gerarden 2012). While there is evidence to support the notion that rebates can be reduced over time (or phased out) as a market is transformed for *certain products*,[1] there is little evidence to support the notion that in most markets for comprehensive energy efficiency in buildings, "financing only" programs can successfully replace broader approaches that combine attractive financing with incentives (e.g. rebates), technical assistance to customers, marketing/education, trade ally partnerships and complementary policies.

The financing gap, to many, seems like a "solvable problem" that can be addressed with politically attractive ideas like public-private partnerships and private sector innovation. The literature on energy efficiency often lists "high first costs" as a key barrier to investment (IEA 2008; Jaffe & Stavins 1994). In our experience examining efficiency programs across the country, lack of financing is seldom the primary reason that efficiency projects do not happen. Financing is only useful once the "product" has been sold to the customer, just as a car loan can only be appealing once you want a car (and then only if there are no better payment options available). Financing cannot address the range of challenges to scaling energy efficiency investment – barriers which include information and hassle costs, split incentives, performance uncertainty, and lack of monetization of public benefits (Golove & Eto 1996, Blumstein et al. 1980). In a world of limited program budgets, program administrators sometimes face a zero- sum choice between allocating funds to supporting financing and allocating funds to approaches designed to overcome a broader set of efficiency barriers. In this paper we explore a set of questions to tease out when financing can be a useful tool, and attempt to highlight some of the limitations of financing to help policy makers and program administrators decide how to allocate resources. These questions are:

- Can financing **increase the leverage** of public funds?
- Can financing **motivate demand** for energy efficiency?
- Can financing **expand access** to energy efficiency?

These questions reflect many of the assumptions made by those promoting energy efficiency financing. We show that in some ways financing *can* do all of these things, but only in certain situations and not always more effectively than alternative uses of public funds. This paper does not provide a prescriptive path for program administrators. Approaches to addressing these complex challenges will vary by market segment and region – and there is a need for innovation in both the public and private sectors. This innovation is likely to change the market dynamics beyond what we describe in this paper. The questions we raise are simply a place to start to consider what financing might offer – and where the limits of financing may lie.

CAN FINANCING INCREASE THE LEVERAGE OF PUBLIC FUNDS?

Current public funding levels are simply not sufficient to pay for a substantial portion of the energy efficiency upgrades necessary to achieve our public policy objectives or capture achievable potential for energy efficiency (Goldman et al. 2010; McKinsey 2009). Financing has been advanced as a tool that can increase the leverage of public monies – that is, increase the level of private investment for each public dollar spent – and potentially lead to energy improvements at a much larger scale than today's activity. Program monies typically support third-party financing in one of two ways[2]:

- **Interest rate buy downs** (IRB) reduce the rate of interest a customer pays below the market rate.
- **Credit enhancements**, typically in the form of loan loss reserves, that reduce a lender's risk in the event of a customer loan default and, in so doing, incentivize lenders to offer more attractive financing products to customers.

We use an example from the residential sector to explore whether financing can increase the leverage of public funds in practice, but the lessons learned apply to the building stock more broadly. Early results from the Energy Upgrade California program show average project costs of about $13,000 for single-family residential energy upgrades and rebates of approximately $2,500 in the San Francisco Bay Area. What would it look like if these rebate funds were transitioned to support financing – and could the level of per-project public funding be *decreased* if it was targeted at supporting financing?

Transitioning the $2,500 in current per project public incentives to support financing would yield an IRB of approximately 5 percent below the market interest rate on a 10 year unsecured term loan. For the Fannie Mae Energy Loan, this implies a post-buy down customer interest rate of approximately 9.7 percent (see Figure 1).[3] Some experts[4] believe that double digit interest rate loans are significant demotivators for households, that very low interest rate loans (e.g. heavily subsidized) can help to sell energy improvements, and that interest rates in the five to nine percent range are enablers rather than drivers of efficiency investment, i.e. if a customer already wants to do efficiency work and doesn't have other financing options, they may take a loan at this rate if they qualify.[5] This suggests that while this offer may allow some customers to do efficiency work when they wouldn't otherwise have had access to attractive capital, it is not likely to increase the demand for efficiency – and it will be difficult to reduce the public cost per project by simply transitioning funds currently spent on rebates to IRBs on loan products at the current market rates offered by capital sources such as Fannie Mae, though there may be cheaper locally-available capital that can be subsidized to much lower rates.

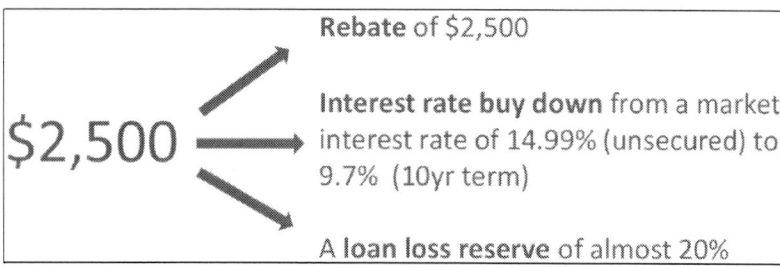

Figure 1. Transitioning From Rebates to Financing.

Figure 2. Increasing Leverage with Credit Enhancements.

Alternatively, this $2,500 could be channeled into a 19% loan loss reserve,[6] which implies that it is through credit enhancements that financing may be effective at increasing leverage. Across the country, there are numerous examples of local lending institutions, typically credit unions, community development financial institutions (CDFIs), and local banks offering single digit interest rate loan products with loan loss reserves of just 5 to 10 percent,[7] well below the 19 percent LLR that $2,500 could achieve. It is possible that a 10 percent loss reserve (costing $1,300 – see Figure 2) plus a rebate of $500 to $1000 would drive customer demand, and have a lower per-project cost than the $2,500 rebate scenario. It is unclear whether this offer would be as or more attractive than the rebate scenario, but it is at least an example where financing can potentially offer some leverage.

However, the question then arises whether credit enhancements can actually deliver this enhanced leverage at scale. In many cases, these local lenders are mission-oriented and see energy efficiency financing as a way to serve their core missions. Some lenders also see energy efficiency financing as a low-cost customer acquisition tool. Compared to their standard product offerings, these lenders approve energy efficiency loan applicants at higher rates, approved loans are funded at higher rates, and borrowers are being cross-sold into other financial product offerings (Zimring 2011b). In other words, local lenders are often subsidizing the interest rate on these loan products as a marketing tool or because it serves their mission.

As this market grows, credit enhancements of three times the expected default level on a loan portfolio may be necessary to reach secondary markets,[8] a step seen by many as a key element of unlocking the billions of dollars of low-cost capital necessary to fund a large scale national investment in efficiency. While it is difficult to use historic default rates on unsecured loans as a guide given recent economic uncertainty, it is reasonable to assume that these non- payment rates may range from the mid single digits to 15 percent for creditworthy consumers, suggesting that credit enhancements of 15 to 45 percent may be necessary to access secondary markets at single digit interest rates. If this is the case, financing is likely to offer similar (or perhaps even less) leverage, at least in the short term, and drive less demand than rebate-driven programs which often cover between a tenth and a third of project costs.[9]

There is often a misperception that, unlike rebates, credit enhancements will last indefinitely and be revolved to support many projects through time. In some instances this may be the case. Whether credit enhancements need to be replenished or not is a function of how large they are relative to loan default rates. Some suggest that loan default rates will be much lower than 5 to 15 percent (Bell, Nadel & Hayes 2011), as energy efficiency lending is fundamentally more secure than lending for other purposes because energy upgrades improve a borrower's cash flow, leaving them with more money to pay back their loans. However, there is reason to doubt, at least with existing programs, that efficiency lending is meaningfully more secure – there is significant variance across the country in actual customer savings and even if savings are realized, there are no promises that borrowers will allocate these funds to repaying the loans. But, to the extent that efficiency lending proves to be more secure (perhaps because it is attached to one's property or utility meter), strong loan performance today may reduce the need for ongoing investment in publicly-funded credit enhancements, and catalyze more attractive and accessible financing products in the future.

Ultimately, it is clear that financing *can* increase program leverage, but whether it *will* increase program leverage remains a complex and open question that is partly a function of whether energy efficiency financing products outperform other lending tools and partly a function of how customer demand might change with a transition from rebate-driven programs to financing, an issue which we discuss in the next section.

CAN FINANCING MOTIVATE DEMAND FOR ENERGY EFFICIENCY?

Energy efficiency programs that have been successful in reaching significant portions of their target markets have typically offered large financial incentives that covered 50 percent or more of the project cost (Fuller et al. 2010). With limited public funding, efficiency programs are tasked with motivating millions of households and businesses to spend thousands of dollars on unfamiliar investments. There are good reasons that people aren't making these improvements, and overcoming these investment barriers is a difficult task at anytime, and even more daunting in an bad economy.

While financing can increase program leverage, the previous section took customer demand as a given. In certain markets, like affordable multifamily housing, financing may indeed be the largest barrier to investment in energy efficiency and affordable financing options can trigger large efficiency investments. In institutional markets, financial innovations such as energy savings performance contracts and performance guarantees have been an important driver of efficiency demand (Satchwell et al. 2010). However, the ESCO business model based on performance contracting (e.g. performance guarantees that savings will be sufficient to pay debt service obligation and third party financing) has had the most success in the institutional sector (e.g., state/local/federal governments, K-12 schools, universities/colleges, and hospitals), which are often the largest, highest credit quality buildings.[10]

However, in most markets, demand – not access to affordable capital – has been the primary barrier to market growth. There is reason for skepticism that most households and businesses considering energy efficiency improvements will be equally or more motivated by attractive financing as they are by today's rebate-driven programs. Even where more affordable financing options are important to overcoming the upfront costs of energy upgrades, low-cost financing, alone, has failed to push people over the edge and motivate wide-scale efficiency investment (Fuller 2009).[11]

In addition, many individuals and institutions already have access to relatively low-cost capital in the form of savings, capital and operating budgets, bonding, home equity lines of credit (HELOCs) or other sources. For these building owners, rebates improve the economics of projects whereas additional financing options – unless heavily subsidized (e.g. zero percent interest rates) – are unlikely to offer substantial value. A homeowner who has access to a five percent home equity line and a 8.99 percent unsecured loan from their credit union will likely take a zero percent interest loan to save money, but what is the public benefit derived from the interest rate subsidy? Would the customer have used their other options and gone through with an energy upgrade in the absence of the zero percent financing? Financing subsidies are often extremely expensive – we need to be open about how much these subsidies cost and what they are achieving relative to alternative uses of scarce public funds. Table 1 shows the typical cost of reducing the market interest rate of an unsecured term loan by 5 percent for a range of loan amounts and terms.

Programs that have successfully achieved relatively high levels of participation by offering financing in lieu of rebates have largely funded improvements like new equipment (e.g. boilers, HVAC systems) in situations where old equipment has failed or needs replacement. Even these programs, which have the advantage of funding improvements familiar and vital to most customers (as opposed to air sealing, insulation, duct sealing, etc), are struggling to achieve scale above one percent of the population annually, and there is an open question about both the additionality of the investments being made and whether the implicit subsidies being allocated to reduce interest rates might be better spent on different types of customer incentives. For example, in Pennsylvania, the Keystone HELP program offers 2.99 percent to 8.99 percent financing to residential customers depending on the comprehensiveness of energy improvements. The program has averaged several thousand projects per year, with loans funded by the state treasury. The PA Treasurer is now struggling to sell this loan pool to investors without offering an enhanced interest rate (e.g. a rate higher than that being paid by program participants) or overcollateralizing the loan pool,[12] both of which increase the cost of offering the program.

In the short term, it is likely that without policies compelling properties to enhance their energy performance, both rebates and financing (and other market development initiatives) will be necessary to scale energy efficiency investment in the building stock. And this seems appropriate – we are not asking people to invest in energy efficiency solely because we want them to save money and be more comfortable. We are motivated to pursue energy upgrades by the range of public benefits that efficiency provides (e.g. reducing the cost of serving energy consumers, easing congestion on the grid, minimization of environmental impacts, equitable access to efficiency opportunities), and those public benefits should be recognized in the form of rebates and other financial incentives.

Table 1. Cost of an Interest Rate Buydown of 5%

Cost of Interest Rate Buydown (14.99% to 9.99%)					
Loan term =	3 years	5 years	7 years	10 years	15 years
Project cost of $3000	$208	$321	$419	$543	$697
Project cost of $6000	$415	$641	$838	$1,086	$1,394
Project cost of $9000	$623	$962	$1,257	$1,628	$2,090
Project cost of $12000	$830	$1,283	$1,677	$2,171	$2,787
Project cost of $15000	$1,038	$1,603	$2,096	$2,714	$3,484
Equivalent rebate level as precent of project cost =	7%	11%	14%	18%	23%

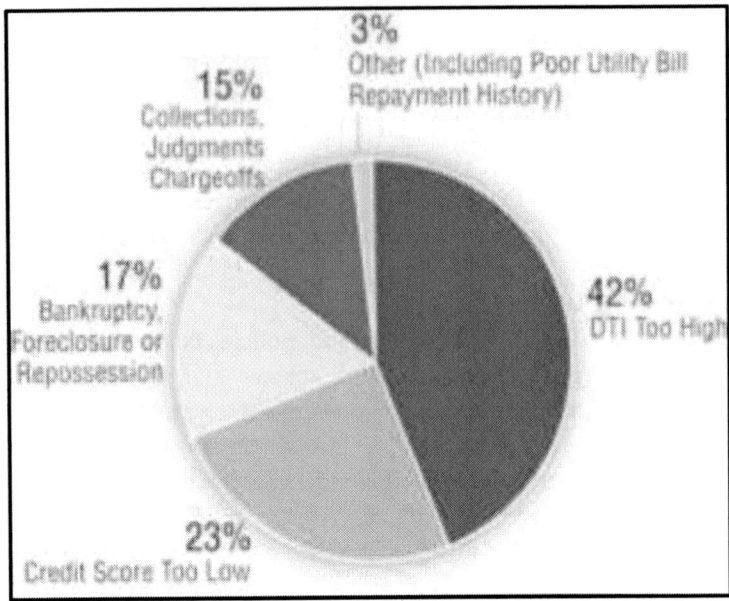

Source: NYSERDA Green Jobs-Green New York Program (November 2010 to October 2011); graphic from Zimring et al. 2011a[13].

Figure 3. Reasons for application rejection in NYSERDA's residential loan program.

CAN FINANCING EXPAND ACCESS TO ENERGY EFFICIENCY?

Once a customer *wants* to invest in efficiency, the question becomes whether they have access to capital. Access to capital varies dramatically across different market segments. In the public and institutional sectors, customers often (though not always) have access to low cost funds through bonding or other sources. If the project is large enough, these customers can work with an energy service company (ESCO) to secure debt with a performance guarantee. For large commercial building and industrial facility owners, access to capital varies widely based on the owner's credit – for example, Class A office buildings typically have access to cheap working capital and many have done basic energy improvements with quick paybacks such as lighting replacements with these funds. In contrast, lower-value commercial and multifamily properties often have little or no access to capital, or may have more urgent uses for their limited debt capacity.

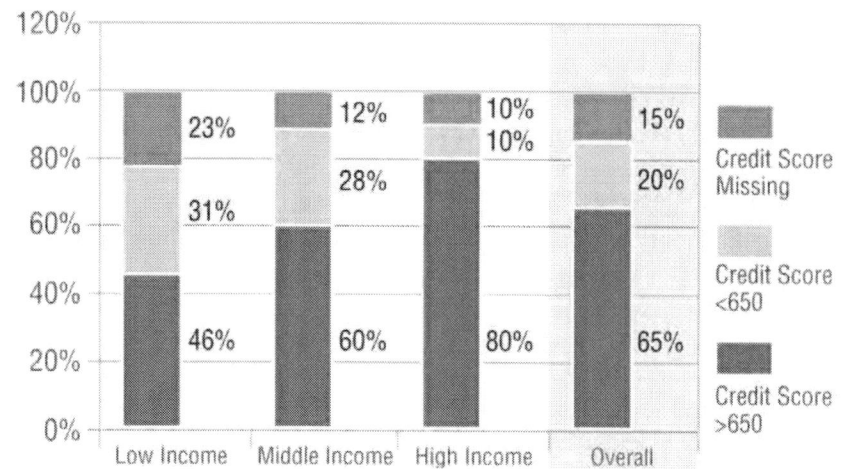

Source: Energy Programs Consortium provided data; graphic from Zimring et al. 2011a.

Figure 4. Homeowner credit scores by income in Q4 2010.

In the residential sector, a significant portion of the population does not qualify for credit, or can only get access to high interest, short term products (Zimring et al. 2011a). Many of the largest energy efficiency loan programs have application decline rates in the 30 to 50 percent range (see Figure 3 for examples of reasons for declines). Household ability to obtain secured financing has declined as housing prices have eroded and lenders have tightened underwriting standards and credit limits. Similar tightening trends are occurring in unsecured lending as personal creditworthiness has weakened and lenders have responded by increasing the minimum credit scores required and reducing the amount of overall credit available to each qualified borrower (Zimring et al. 2011a). Access to capital can sometimes (but not always) be correlated, perhaps understandably, to income (see Figure 4).

In general, those individuals and organizations who most need access to external sources of capital to pay for energy improvements are more often rejected from financing programs because they are simply deemed less creditworthy by current underwriting metrics. If programs only offer financing to those who *already* have access to affordable capital, there may be minimal additional benefit from public support for these programs. Efficiency program administrators need to ask themselves if the products they create are filling an unmet need.

There are ways to make capital more accessible and affordable to underserved markets, with prudent safeguards. Alternative underwriting may be a way to expand access to credit, on the margins, to those who don't currently qualify for existing products – though we need more experimentation in this area. For example, the New York State Energy Research and Development Authority (NYSERDA) is using utility bill payment history to assess credit quality, which has led to the approvals of an additional 5 percent of applicants.[14] There is not enough history to adequately assess the performance of these loans, but the early results are encouraging. Expanding access to credit may also simply require more public investment – larger loan loss reserves and other credit enhancements may be necessary to serve many market segments, and this use of funds should be compared to other options in terms of impact. These programs may also require this investment on an ongoing basis – unless we can prove that lending for energy efficiency is fundamentally more secure than other financial tools, deserving lower rates and expanded access, that certain market segments are more creditworthy than current metrics imply, or that new products like on-bill finance will significantly improve repayment rates. Program administrators need to collect and analyze the data required over many years to make this case, and they need to be clear up front whether they are trying to prove that new or different products work better than conventional products or metrics of credit, or if they are simply aiming to subsidize access to credit for less creditworthy customers.

It is also important to note that underwriting criteria exist for a reason – to ensure that those that get access to financing are willing and able to make the payments. There can be negative consequences to promoting loans to particularly vulnerable segments of the population. Cae needs to taken with regard to who is given access to credit and what claims are being made about the benefits of energy improvements. It is likely that financing will never serve *all* customers, nor should it.

CONCLUSIONS

Financing is an appealing concept when efficiency program budgets are a small fraction of the overall level of efficiency investment needed to achieve our public policy goals – but that does not mean financing is always the best solution for increasing the uptake of efficiency measures, and it is certainly not the *only* solution. We began this paper with three questions about financing: Can financing **increase the leverage** of public funds, **motivate demand** for energy efficiency, and **expand access** to efficiency? We have shown that financing can, in some cases, do all of these things. With a loss reserve, financing can increase the leverage of public dollars, but in most cases it is not able to drive demand to the same degree as direct incentives like rebates, and cannot be expected to replace other incentives in the current marketplace. We have also argued that subsidized financing for those who already have access to capital may be a poor use of public funds, and that increasing access for those who are underserved by existing financial products is possible and important, but will likely require ongoing subsidy.

This is not to say that financing is unimportant. Once a customer wants to invest in efficiency, financing must be available for those who don't have alternative affordable options for payment. We also should encourage participants to pay for as much of the efficiency investment as possible to avoid the political consequences of short term rate impacts and to help spread limited public dollars as far as possible. We just need to be clear about what financing can accomplish, and not assume that it is the single solution to the many challenges to scaling energy efficiency. Scaling efficiency requires selling the product much more effectively to customers – making it simple, seamless, attractive, and affordable – and perhaps more importantly, paying for efficiency – often the lowest cost energy resource – as a resource on par with other energy supply options to make sure we aren't spending more money than needed to meet energy demand.

REFERENCES

Bell, Catherine J., Steven Nadel & Sara Hayes. (2011). "On-Bill Financing for Energy Efficiency Improvements: A Review of Current Program Challenges, Opportunities, and Best Practices." Report Number E118. Washington, DC: ACEEE.

Blumstein, Carl, Krieg, B., Schipper, L. & York, C. (1980). "Overcoming social and institutional barriers to energy conservation." Energy, Volume 5, Issue 4, April, 355-371. ISSN 0360-5442.

Brown, Matthew & Beth Conover. (2009). "Recent Innovations in Financing for Clean Energy." Boulder, Colorado: Southwest Energy Efficiency Project.

Golove, William & Eto, J. (1996). "Market Barriers to Energy Efficiency: A Critical Reappraisal of the Rationale for Public Policies to Promote

Energy Efficiency". Berkeley, Calif.: Lawrence Berkeley National Laboratory, Environment Energy Technologies Division. LBNL-38059.

Fuller, Merrian, Cathy Kunkel, Mark Zimring, Ian Hoffman, Katie L. Soroye, & Charles Goldman. (2010). "Driving Demand for Home Energy Improvements: Motivating Residential Customers to Invest in Comprehensive Upgrades that Eliminate Energy Waste, Avoid High Bills, and Spur the Economy." Berkeley, Calif.: Lawrence Berkeley National Laboratory, Environmental Energy Technologies Division. LBNL-3960E.

Fuller, Merrian. (2009). "Enabling Investments in Energy Efficiency: A study of energy efficiency programs that reduce first-cost barriers in the residential sector." Report prepared for Efficiency Vermont and California Institute for Energy and the Environment.

Goldman C., Fuller, M., Stuart, E., Peters, J., McRae, M., Albers, N., Lutzenhiser S. & Spahic, M. (2010). "Energy Efficiency Services Sector: Workforce Size and Expectations for Growth." Berkeley, Calif.: Lawrence Berkeley National Laboratory, Environmental Energy Technologies Division. LBNL-3987E.

[IEA] International Energy Agency. (2008). "Promoting Energy Efficiency Investments." ISBN 978-92-64-04214-8.

Jaffe, Adam B. & Robert N. Stavins. (1994). "The Energy-Efficiency Gap: What does it mean?" Energy Policy (Volume 22, Number 10), 804-810.

McKinsey & Company. (2009). "Unlocking Energy Efficiency in the US Economy." New York and London.

Palmer, Karen, Margaret Walls & Todd Gerarden. (2012). "Borrowing to Save Energy: An Assessment of Energy-Efficiency Financing Programs." Resources for the Future.

Satchwell, A., Goldman, C., Larsen, P., Gilligan, D. & Singer, T. (2010). "A Survey of the U.S. ESCO Industry: Market Growth and Development from 2008 to 2011." Berkeley, Calif.: Lawrence Berkeley National Laboratory, Environmental Energy Technologies Division. LBNL-3479E.

Zimring, Mark, Merrian Borgeson, Ian Hoffman, Charles Goldman, Elizbeth Stuart, Annika Todd & Megan Billingsley. (2011a). "Delivering Energy Efficiency to Middle Income Single Family Households." Berkeley, Calif.: Lawrence Berkeley National Laboratory, Environmental Energy Technologies Division. LBNL-5244E.

Zimring, Mark. (2011b). "Austin's Home Performance with ENERGY STAR Program: Making a Compelling Offer to a Financial Institution Partner." Berkeley, Calif.: Lawrence Berkeley National Laboratory, Environmental Energy Technologies Division. LBNL-4396E.

End Notes

[1] See, for example, the work of the Northwest Energy Efficiency Alliance (http://neea.org) and proceedings of ACEEE's National Symposium on Market Transformation (http://www.aceee.org/conferences/mt/past).

[2] Direct program lending with public funds does not "leverage public monies" and is not addressed in this paper.

[3] All IRB calculations in this report are from the Department of Energy's LLR and IRB Allocation and Expenditure calculator. The base interest rate on the Fannie Mae Loan ranges from 14.99 to15.99 percent.

[4] Based on the authors' conversations with a wide range of contacts from the financing industry and current EE financing program administrators.

[5] In this section, we take demand as constant – the question of how financing impacts demand is explicitly addressed in the next section.

[6] LLRs reduce lender risk by providing first loss protection in the event of loan defaults. For example, a 5 percent LLR allows a private lender to recover up to 5 percent of its portfolio of loans from the LLR. A $20 million fund of private capital would need a $1 million public LLR (5 percent coverage), leveraging each public dollar 20 to 1. On any single loan default, the LLR typically pays only a percent of the loss (often 80 percent) to ensure the lender is incentivized to originate loans responsibly.

[7] Examples include programs in Michigan (MichiganSaves), Oregon (Clean Energy Works), California (EmpowerSBC), and others.

[8] Alfred Griffin, Citigroup, CPUC Financing Workshop Panel Discussion. February 9, 2012.

[9] Default rates vary dramatically across market segments and financial product types – for some markets, LLRs may be the lowest-cost tool available to program implementers.

[10] Several companies are now offering energy efficiency as a service in the investment grade commercial sector in which building owners pay for energy improvements through time with the savings through time without taking on debt.

[11] Though, contractors and program managers do suggest that when low-cost financing is available, it often encourages larger projects (and deeper energy savings) for customers planning to make improvements.

[12] Overcollateraliztion involves offering investors a pool of loans with total value greater than the value which they are being asked to pay.

[13] DTI is an acronym for Debt-To-Income ratio, a common underwriting metric designed to ensure that borrowers have sufficient cash flow to make principal and interest payments on loans. Typically, maximum DTI's are 40-50 percent (50 percent in NYSERDA's case).

[14] Jeff Pitkin, NYSERDA Treasurer, email correspondence May 1, 2012.

INDEX

A

abatement, 76
access, vii, viii, 9, 10, 12, 13, 15, 16, 17, 19, 20, 21, 26, 35, 42, 47, 48, 54, 57, 59, 60, 61, 62, 63, 64, 65, 66, 75, 76, 77, 79, 80, 82, 83, 84, 85, 86, 87
accessibility, 41
accounting, 47
administrators, vii, 1, 2, 3, 4, 6, 8, 14, 15, 16, 17, 18, 22, 23, 24, 25, 27, 28, 29, 32, 35, 37, 41, 44, 48, 49, 69, 76, 77, 78, 85, 86, 89
aesthetics, 28
age, 37, 38, 43
agencies, 68
aggregation, 2, 8, 14
American Recovery and Reinvestment Act, 48
annual rate, 59
assessment, 4, 9, 11, 35, 69
assets, 55, 73
asymmetry, 12
attachment, 74
aversion, 21

B

balance sheet, 9, 10, 47, 70
bankruptcy, 61, 72, 73
banks, 48, 80
barriers, vii, 2, 4, 6, 8, 9, 11, 14, 15, 20, 45, 61, 73, 77, 81, 87, 88
base, 20, 89
behaviors, 27
benefits, viii, 6, 11, 12, 15, 21, 26, 28, 48, 49, 51, 53, 62, 76, 77, 83, 86
bias, 30, 33, 38, 50
boilers, 82
bonding, 82, 84
bonds, 74
borrowers, viii, 54, 64, 72, 73, 74, 80, 81, 89
business model, 82
businesses, 17, 19, 49, 76, 81, 82
buyers, 25

C

capital markets, 11
carbon, 76
cash, 12, 62, 68, 71, 72, 76, 81, 89
cash flow, 12, 62, 68, 72, 76, 81, 89
Census, 43, 51, 71
challenges, viii, 28, 48, 54, 77, 78, 87
cities, 56, 63
City, 56
classes, 6, 15, 33, 41, 42, 44
clean energy, 47, 67
climate, 30, 34, 74, 76

collateral, 12
colleges, 82
commercial, 48, 84, 89
communities, 36
community, 48, 69, 80
competition, 23
conformity, 74
consent, 10
consumer investment, vii, 2, 4
consumer protection, 74
consumers, 2, 6, 7, 8, 9, 10, 11, 13, 14, 15, 17, 19, 20, 21, 22, 23, 24, 25, 26, 28, 29, 30, 32, 33, 34, 35, 36, 37, 38, 39, 40, 41, 42, 43, 44, 45, 48, 49, 50, 63, 76, 80, 83
control group, 28, 30, 49
conversations, 89
cooling, 43
correlation, 51
cost, vii, 2, 4, 8, 10, 11, 14, 15, 16, 17, 18, 19, 24, 27, 32, 41, 45, 47, 48, 49, 54, 57, 62, 63, 66, 70, 74, 76, 77, 79, 80, 81, 82, 83, 84, 87, 88, 89
creditworthiness, 48, 58, 59, 60, 61, 64, 65, 66, 69, 72, 73, 85
culture, 51
customer-funded efficiency programs, vii, 1
customers, 47, 48, 49, 50, 51, 60, 63, 64, 70, 73, 74, 76, 77, 78, 79, 82, 84, 86, 87, 89

D

data analysis, 32
data collection, 26, 27, 43, 48
data set, 13, 26, 70
database, 35
debt service, 20, 21, 72, 73, 82
deduction, 68, 74
delinquency, 59, 60, 62, 67, 72
Department of Energy, 1, 26, 48, 71, 89
depression, 71
depth, 17, 21, 28
discontinuity, 32, 50
diversification, 11
diversity, 26
drawing, 24

E

earnings, 57
economic development, 68
economics, 17, 35, 82
economies of scale, 33
education, 77
elasticity of demand, 20
electricity, 11
emission, 7
employees, 68, 74
employers, 74
employment, 38
encouragement, 50
energy conservation, 87
energy efficiency, vii, viii, 1, 3, 4, 5, 8, 9, 11, 12, 13, 14, 17, 18, 28, 29, 32, 35, 36, 40, 45, 48, 49, 50, 53, 54, 57, 61, 62, 64, 66, 68, 69, 74, 76, 77, 78, 80, 81, 82, 83, 85, 86, 87, 88, 89
Energy Efficient Mortgage, 47
energy prices, viii, 53, 76
energy supply, 87
environment, 16, 19, 76
environmental impact, 11, 76, 83
equipment, 21, 68, 82
equity, viii, 19, 28, 47, 54, 56, 60, 68, 69, 71, 82
evidence, 13, 17, 23, 25, 77
expenditures, 58, 71
experimental design, 6, 24, 27, 28, 29, 30, 32, 33, 36, 37, 39, 40, 42, 43, 44, 49, 50, 51
exposure, viii, 53
externalities, 49

F

fairness, 35
families, 56, 57, 68, 74
Fannie Mae, 74, 79, 89
federal government, 82
Federal Reserve, 61, 70, 71, 72
Federal Reserve Board, 61, 70, 71, 72

Index

financial, 2, 6, 7, 8, 9, 10, 11, 12, 13, 14, 16, 19, 20, 21, 22, 23, 25, 26, 27, 28, 29, 41, 44, 47, 48, 49, 55, 56, 57, 63, 67, 69, 70, 74, 80, 81, 83, 86, 87, 89
financial crisis, 55
financial incentives, 14, 57, 81, 83
financial innovation, 81
financial institutions, 6, 12, 13, 14, 22, 26, 44, 48, 67, 70, 74, 80
financial markets, 2, 10, 25, 26, 27
flexibility, 48, 65
focus groups, 25
foreclosure, 47, 61, 72
formal education, 47
full employment, 71
funding, vii, 1, 3, 8, 14, 46, 47, 67, 70, 78, 79, 81, 82
funds, vii, viii, 1, 4, 5, 7, 9, 11, 14, 16, 48, 49, 63, 68, 75, 77, 78, 79, 81, 82, 84, 86, 89

G

government funds, 47
governments, 76
greenhouse, viii, 53
greenhouse gas(es), viii, 53
growth, 57, 82
guidance, 4
guidelines, 74

H

health, viii, 28, 53, 68
hedging, 67
historical data, 48
history, 65, 66, 73, 86
home value, 56, 71
homeowners, 50, 54, 56, 57, 59, 62
homes, viii, 53, 54, 56, 68, 69
household income, 71, 72
housing, 54, 55, 56, 58, 68, 71, 81, 85
HUD, 47
hypothesis, 17

I

ideal, 29
IEA, 7, 77, 88
improvements, vii, viii, 2, 4, 6, 7, 8, 9, 10, 11, 12, 13, 14, 17, 19, 20, 21, 22, 23, 26, 27, 28, 29, 38, 41, 44, 47, 48, 49, 53, 54, 57, 58, 59, 61, 62, 66, 67, 68, 71, 76, 77, 78, 79, 81, 82, 84, 85, 86, 89
income, viii, 37, 38, 41, 42, 43, 51, 53, 54, 55, 56, 57, 58, 60, 61, 62, 63, 64, 67, 68, 69, 71, 72, 73, 74, 85
increased access, 65
individuals, 49, 62, 76, 82, 85
industry, 89
information technology, 49
infrastructure, 2, 14, 16
institutions, 12, 13, 22, 48, 49, 70, 76, 80, 82
insulation, 41, 82
integrity, viii, 53
interest rates, 2, 12, 13, 19, 20, 38, 51, 63, 72, 79, 80, 82, 83
International Energy Agency, 46, 88
intervention, 5, 7, 13, 14, 49
investment(s), vii, viii, 1, 2, 3, 4, 6, 9, 11, 14, 15, 24, 30, 38, 46, 47, 48, 49, 54, 57, 58, 75, 76, 77, 79, 80, 81, 82, 83, 86, 87, 89
investors, 8, 9, 13, 25, 47, 48, 70, 83, 89
issues, 2, 6, 32, 38, 60, 62, 69

J

justification, 9

L

labeling, 15
lead, 9, 11, 12, 16, 30, 38, 68, 74, 78
legislation, 23, 69
lending, 11, 14, 58, 64, 67, 68, 72, 73, 74, 80, 81, 85, 86, 89
lifetime, 7, 62

liquidity, 48
loan principal, 74
loans, 13, 14, 33, 40, 48, 58, 59, 62, 63, 64, 65, 66, 67, 68, 69, 70, 71, 72, 73, 74, 79, 80, 81, 83, 86, 89
local government, 7, 48

M

magnitude, 37
majority, 56, 62
market access, 70
market failure, 7, 8, 14
market penetration, 4, 17
market segment, 2, 8, 9, 11, 18, 42, 43, 78, 84, 86, 89
marketing, 13, 47, 49, 74, 77, 80
marketplace, viii, 72, 75, 87
matter, 16
median, 56, 71
meter, 10, 67, 74, 81
mission(s), 74, 80
models, viii, 9, 22, 54, 67, 72
modifications, 8
motivation, 58, 73

N

negative consequences, 62, 86
negative equity, 56
neutral, 29

O

Oklahoma, 48
opportunities, viii, 48, 49, 54, 76, 83
ownership, 23, 43

P

participants, 17, 18, 25, 28, 30, 34, 49, 63, 65, 83, 87
pathways, 14
payback period, 58

payroll, 68
percentile, 71
personal accounts, 74
personal control, 74
pessimism, 57
pitch, 34, 35
policy, 4, 9, 15, 21, 23, 24, 29, 46, 47, 74, 77
policy makers, 77
policymakers, vii, 1, 2, 3, 4, 6, 8, 14, 15, 16, 21, 24, 25, 27, 32, 47, 51, 70, 76
pollutants, viii, 53
pools, 13, 14, 25, 26, 27, 48, 69, 70
population, 56, 62, 82, 85, 86
portfolio, 4, 73, 80, 89
potential benefits, 11
power system costs, viii, 53
private benefits, 6, 49
private investment, 78
private sector, 19, 47, 77, 78
product eligibility, 63
product performance, 12, 28, 29
profitability, 48
program administration, 49
program features, 27
program outcomes, 50, 51
program staff, 24
project, 9, 10, 12, 18, 21, 25, 26, 27, 28, 30, 41, 47, 78, 79, 80, 81, 83, 84
property taxes, 10, 69
protection, 73, 89
public investment, 86
public policy, viii, 75, 76, 78, 86
public support, 85
public-private partnerships, 77

Q

qualitative research, 26
quality assurance, 67

R

random assignment, 28, 35, 49
real estate, 19, 74

reality, 76
recession, 54, 55, 56, 57, 59, 60, 71
recommendations, 50
regression, 32, 50
rejection, 58, 61, 84
rent, 10
requirements, 61
reserves, 23, 48, 63, 73, 78, 80, 86
Residential, v, 46, 48, 49, 53, 69, 88
resources, vii, 2, 4, 7, 20, 21, 23, 24, 25, 49, 77
response, 29
revenue, 67
risk(s), viii, 9, 11, 12, 13, 15, 25, 30, 37, 48, 54, 60, 62, 63, 66, 67, 69, 70, 72, 73, 78, 89
risk assessment, 12, 48, 66
rules, 23

S

safety, 28, 68
savings, vii, 1, 2, 3, 4, 6, 7, 8, 9, 11, 12, 13, 14, 15, 16, 18, 20, 21, 22, 26, 28, 30, 40, 41, 49, 54, 57, 62, 67, 68, 76, 81, 82, 89
scaling, viii, 75, 77, 87
school, 22, 82
scope, 30, 34
seasonal changes, 37
security, 9, 19, 21, 47, 69, 74
self-employed, 73
service provider, 6, 15
services, 27, 68, 76
small businesses, 10
solution, viii, 75, 76, 86, 87
spending, 42, 71, 87
stakeholders, 6, 24, 47
standardization, 8, 13, 25, 26, 27, 48
state(s), vii, 1, 3, 6, 7, 22, 24, 25, 36, 46, 47, 48, 56, 57, 71, 74, 76, 82, 83
state regulators, 76
stock, 78
stress, 56
structure, 21, 43, 63, 74
subsidy, viii, 75, 82, 87

T

TAP, 71
target, 6, 11, 15, 20, 27, 35, 41, 44, 63, 81
tariff, 21
tax rates, 72
taxes, 72, 73
technical assistance, 2, 15, 17, 27, 68, 77
techniques, 6, 24, 25, 27, 30, 33, 34, 49, 50
technology, 47
tenants, 10, 23
territory, 36
testing, 25, 29, 35, 49, 51
trade, 77
training, 15, 27
transaction costs, 20, 77
transactions, 13, 74
transformation, 48
treatment, 9, 28, 30, 32, 33, 34, 35, 44, 47, 49
trial, 28

U

underwriting, viii, 2, 9, 11, 12, 13, 20, 47, 48, 54, 58, 59, 61, 63, 64, 65, 66, 69, 72, 73, 74, 85, 86, 89
unemployment rate, 57
unions, 48, 68, 70, 80
unique features, 48
United, 55
United States, 55
universities, 82
up-front costs, 7, 17
urban, 36
utility bill payer, vii, 1, 3, 4, 5, 7, 8, 9, 11, 12, 13, 14, 16, 24, 47

V

vacuum, 40
variables, 41
vehicles, 13, 19

W

wages, 57
Washington, 68, 74, 87
wealth, 54, 55, 56, 71
Wisconsin, 64

workers, 57, 71
workforce, 15, 47, 76

Y

yield, 12, 27, 43, 49, 50, 76, 79